Fetal Cells and Fetal DNA in Maternal Blood

11th Fetal Cell Workshop, Basel, April 15, 2000

Fetal Cells and Fetal DNA in Maternal Blood

New Developments for a New Millennium

Editors

Sinuhe Hahn *Basel*
Wolfgang Holzgreve *Basel*

22 figures, 3 in color and 16 tables, 2001

KARGER

Basel · Freiburg · Paris · London · New York · New Delhi · Bangkok · Singapore · Tokyo · Sydney

Sinuhe Hahn, PhD
Laboratory Chief Prenatal Medicine
Department of Obstetrics and Gynecology
University of Basel, Switzerland

Wolfgang Holzgreve, MD, MS, Dr. h.c.
Professor and Chairman
Department of Obstetrics and Gynecology
University of Basel, Switzerland

Library of Congress Cataloging-in-Publication Data

Fetal Cell Workshop (11th : 2000 : Basel, Switzerland)
Fetal cells and fetal DNA in maternal blood : new developments for a new millennium / 11th Fetal Cell Workshop, Basel, April 15, 2000 ; editors, Sinuhe Hahn, Wolfgang Holzgreve.
p. ; cm.
Includes bibliographical references and indexes.
ISBN 3805572220
1. Fetal cells from maternal blood–Congresses. 2. Fetal blood–Congresses. 3. Prenatal diagnosis–Congresses. 4. Fetus–Diseases–Congresses. I. Hahn, Sinuhe. II. Holzgreve, Wolfgang. III. Title.
[DNLM: 1. Prenatal Diagnosis–methods–Congresses. 2. Cell Separation–methods–Congresses. 3. DNA–blood–Congresses. 4. Fetal Diseases–diagnosis–Congresses. 5. Fetus–cytology–Congresses. 6. Polymerase Chain Reaction–Congresses. 7. Pregnancy–blood–Congresses. WQ 209 F41917f2001]
RG628.3.F45 F48 2000
618.3′2075–dc21

2001022537

Bibliographic Indices. This publication is listed in bibliographic services.

Drug Dosage. The authors and the publisher have exerted every effort to ensure that drug selection and dosage set forth in this text are in accord with current recommendations and practice at the time of publication. However, in view of ongoing research, changes in government regulations, and the constant flow of information relating to drug therapy and drug reactions, the reader is urged to check the package insert for each drug for any change in indications and dosage and for added warnings and precautions. This is particularly important when the recommended agent is a new and/or infrequently employed drug.

www.karger.com
Printed in Switzerland on acid-free paper by Reinhardt Druck, Basel
ISBN 3–8055–7222–0

Contents

Preface

It is an impossible task to stand on the brink of a new century and attempt to gaze into the future to see what new developments we will be faced with, not only in our daily lives, but especially those affecting our clinical and research decisions. As we approached the end of the last century we witnessed dramatic changes in population demographics in the developed world, which have far-reaching consequences for health services. This is not only evident by an increasingly ageing population, but also by an increase in the number of women who become pregnant at a more advanced maternal age, many of whom are opting to have a single child. This situation has placed obstetricians and genetic counselors in an awkward position, in that these women have an increased risk of having a fetus with a chromosomal aberration, and many are reluctant to expose this long-desired child to the risks associated with an invasive prenatal diagnostic procedure. A further concern is the high false-positive rate for the detection of fetal aneuploidies associated with current non-invasive methods, such as ultrasound and/or serum analyte screening, as this leads to a large number of unnecessary invasive procedures which are coupled with considerable costs for an already burdened health care system.

It is very clear that new developments will be necessary to meet these challenges. In this volume we have collected articles from many of the leading researchers who are in the process of developing new accurate risk-free prenatal diagnostic techniques using fetal cells isolated from the maternal circulation, or the more recent development of examining cell free fetal DNA in maternal plasma or serum by PCR. Even though many hopes have been raised, dashed and are being raised again, many in the field are confident that we are indeed approaching that elusive goal, of being able to perform an accurate prenatal diagnosis using non-invasive techniques, such as the enrichment of fetal cells from the blood of pregnant women or by the analysis of circulatory fetal DNA in maternal plasma. Evidence for this renewed optimism was readily shown by the increasing number of groups entering the hunt for fetal cells in maternal blood who were present at the 11th Fetal Cell Workshop in Basel, Switzerland. As organizers of this meeting, it was a particular pleasure for us to welcome speakers and visitors from as far afield as Israel, Finland, and the Czech Republic, apart from the more traditional core from the UK, Germany, the Netherlands, France and Spain. We were also delighted that a very large proportion of these groups were willing to contribute to this volume, which we hope will serve as an important guide with regard to the current state of the art and give a hint of the directions in which the field is heading.

Sinuhe Hahn, Basel
Wolfgang Holzgreve, Basel

Hahn S, Holzgreve W (eds): Fetal Cells and Fetal DNA in Maternal Blood. New Developments for a New Millennium. 11th Fetal Cell Workshop, Basel, 2000. Basel, Karger, 2001, pp 1–7

Circulating DNA in Plasma or Serum

Philippe Anker, Maurice Stroun

Department of Plant Biochemistry and Physiology, University of Geneva, Switzerland

In 1948 Mandel and Métais [1] identified nucleic acids in human blood plasma. Using a perchloric acid precipitation method, they found RNA and DNA in the plasma of both normal subjects and those with various diseases and, interestingly, in the plasma of pregnant women. This early work on plasma and serum nucleic acids appears to have been largely forgotten, and it was not until 1966 that interest in circulating DNA reawakened with the discovery of DNA in the serum of patients with systemic lupus erythematosus (SLE) [2]. Further work showed that DNA was not only found in the plasma of patients with SLE, but was also present in patients with rheumatoid arthritis, glomerulonephritis, pancreatitis, cholelithiasis, inflammatory bowel disease, peptic ulcer disease, hepatitis and esophagitis [3–6].

DNA Is Found in Increased Amounts in Plasma/Serum of Cancer Patients

Leon et al. [7] found that cancer patients tended to have even higher levels of circulating DNA than those with non-malignant diseases. Significantly greater amounts of DNA were found in the serum of patients with metastases compared to localized disease. Furthermore, DNA levels decreased by up to 90% following radiotherapy, especially in lymphoma, lung, ovary, uterus and cervical tumors. In contrast, persistently high or increasing DNA levels were associated with a lack of response to treatment.

Working with various malignancies (leukemia, lymphoma, lung, breast and gastrointestinal tumors) and after having extracted and purified the DNA from the plasma, it was found that detectable amounts of circulating DNA were found only in patients with advanced malignancies bearing a large tumor cell burden [8].

Further studies by researchers working with lung cancer patients found a correlation between circulating levels of DNA and prognosis [9, 10].

Tumor DNA in the Plasma/Serum of Cancer Patients: Biophysical Properties

In the results showing increased levels of plasma DNA in cancer patients it was not determined whether the circulating DNA was released from activated lymphocytes reacting towards the disease as some data from SLE studies might suggest [5] or from the tumor cells themselves. One way to solve the problem was to take advantage of differences between the properties of the DNA of malignant and normal cells, and to find out whether the DNA isolated from the plasma of cancer patients presented some of the characteristics detected in the DNA of malignant cells. In 1989 it was shown that this plasma DNA from cancer patients shared some biophysical properties (decreased strand stability) common to the DNA of cancer cells and hence was of tumoral origin [11]. This approach was interesting but it required microgram amounts of DNA. Luckily PCR techniques became available allowing oncogene mutation detection in minute quantities.

Plasma DNA N-ras Mutations and Rearranged Ig Heavy Chain DNA in Hematological Malignancies

The first mutational analyses performed on plasma DNA concentrated on the K-ras and N-ras oncogenes [12–17].

N-ras mutations have been found in DNA extracted from the bone marrow of patients with myelodysplastic syndrome and acute myelogenous leukemia. These alterations have also been found in the plasma, leukocytes and bone marrow of such patients [14]. In patients with N-ras alterations, mutant DNA was always present in plasma DNA, though sometimes absent in the DNA of peripheral blood cells or bone marrow. This indicates that a single bone marrow biopsy or aspiration does not necessarily contain all the malignant clones involved in the disease.

Follow-Up of Hematological Malignancies by Analysis of Markers in Circulating DNA

The second work on hematological disorders was done on B-cell malignancies where rearranged Ig heavy chain DNA was detected in plasma or serum samples of patients [18]. Tumor-derived clonal CDRIII DNA was found in serum or plasma in 47% of 110 patients with non-Hodgkin's lymphoma or acute B-precursor lymphoblastic leukemia. Follow-up showed a close correlation between persisting tumor-derived plasma/serum DNA and resistant disease or early relapse.

K-ras Mutations in Circulating DNA of Colorectal Cancer Patients

To our knowledge seven publications [12, 13, 15] report the presence of K-ras mutations in the circulating DNA corresponding to the mutation found in the tumor. In one of these studies, the authors have been able to follow-up a number of patients after surgery [15]. Postoperative K-ras mutations were found in the serum of patients who developed a relapse. The plasma mutation detection predated the clinical recurrence by 1 year in 1 case.

K-ras Mutations in Circulating DNA of Pancreatic Cancer Patients

The K-ras gene is mutated in approximately 90% of pancreatic adenocarcinomas, suggesting that a comprehensive analysis of many genes would be unnecessary to detect the majority of cases. Five studies have therefore concentrated on detecting this gene in pancreatic cancer [12, 13, 16, 17].

In one study [16], plasma DNA was isolated from 21 pancreatic cancer patients and K-ras alterations were detected in the plasma of 17 patients (81%). In cases for whom both plasma and pancreatic tissue were available, DNA mutations were similar in corresponding plasma and tissue samples. Plasma DNA alterations were found 5–14 months before clinical diagnosis in 4 patients who had been diagnosed as suffering from pancreatitis showing again that circulating tumor DNA may be an early event in oncogenesis. Mutant DNA was not found in the plasma of 3 patients with chronic pancreatitis who did not develop a carcinoma.

A follow-up study of patients with pancreatic adenocarcinoma also resulted in plasma DNA alterations being detected in a high proportion of cases in whom a K-ras mutation was found in the tumor tissue [17]. Treatment resulted in the disappearance of K-ras gene mutations in plasma DNA in 6 of 9 (67%) patients.

Three patients with a persistently positive K-ras gene mutation in pre- and post-treatment plasma samples showed early recurrence or a progressive disease.

Microsatellite Alterations in Circulating DNA of Cancer Patients

All tumors, however, do not have high mutation rates on easily testable hot spots. This is the reason why several groups looked for microsatellite alterations in the plasma/serum DNA of cancer patients. Microsatellite DNA is composed of simple repeats of unknown function. It is unstable in cancer cells and subject to alterations which appear as new alleles, allele expansion or loss of heterozygosity. Microsatellite DNA alterations are part of neoplastic progression and they may serve as clonal markers.

We initiated a study to detect microsatellite alterations in paired samples of plasma and tumor DNA from patients with small cell lung carcinoma compared to the same repeat sequences of normal cells from the same patients. A microsatellite alteration was present in 16 of 21 (76%) SCLC tumors and in 15 of 21 (71%) corresponding plasma samples [19].

Microsatellite alterations were also found in the circulating DNA of head and neck cancer patients [20]. In this study, positive serum samples seemed to be related to patients with advanced disease, suggesting that they may prove useful as prognostic factors.

Microsatellite alterations have also been detected in the circulating DNA of patients suffering from a variety of malignancies: non-small cell lung [12, 13, 21], renal [22], bladder [23], breast [24], HNPCC and sporadic colon [25], ovarian cancers [26] and melanoma [27].

Other Approaches to Cancer Detection or Follow-Up through Circulating DNA

Other cancer-related markers have been reported in the circulating DNA of cancer patients such as EBV sequences [12, 13, 28] and promoter hypermethylation of several genes particularly p16 [12, 13, 29–31]. It is known that p16 mutations are rare but promoter hypermethylation resulting in gene inactivation is found to be common in a variety of different primary tumors. New markers are available since multiple genes are now known to be somatically inactivated by promoter hypermethylation. Twenty-two patients with non-small cell lung cancer were tested [29], searching for promoter hypermethylation of the tumor-suppressor gene p16, the putative metastasis-suppressor gene death-associated protein kinase, the detoxification gene glutathione S-transferase P1, and the DNA repair gene O6-methylguanine-DNA-methyltransferase. Aberrant methylation of at least one of these genes was detected in 15 of 22 (68%) non-small cell lung cancer tumors but not in any paired normal lung tissue. In these primary tumors with methylation, 11 of 15 (73%) samples also had abnormal methylated DNA in the matched serum samples. Similarly, in the study of liver cancer [30], p16 methylation was found in the plasma/serum samples of 81% (13/16) of patients presenting p16 methylation in their tumor. Hypermethylation of p16 was also reported in tumor and plasma DNA of a series of breast [24] and head and neck patients [31].

Medical Implications of Circulating DNA

The data about this novel approach for cancer detection, prognosis and follow-up are still preliminary from the clinical point of view. What are the implications of plasma DNA for clinical medicine? Firstly, the possibility of a noninvasive plasma DNA test for cancer diagnosis is challenging. Secondly, the association between mutant plasma DNA and tumor stage, the presence of residual disease and outcome suggests that it might potentially serve as a prognostic marker in the absence of clinically detectable metastases. Follow-up tests might be valuable to assess the response to therapy, identifying residual disease and recurrence at an asymptomatic and early stage.

Acknowledements

The authors acknowledge the generous funding by La Ligue Suisse contre le Cancer.

References

1 Mandel P, Métais P: Les acides nucléiques du plasma sanguin chez l'Homme. CR Acad Sci Paris 1948;142:241–243.
2 Tan EM, Schur PH, Carr RI, Kunkel HG: Deoxyribonucleic acid (DNA) and antibodies to DNA in the serum of patients with systemic lupus erythematosus. J Clin Invest 1966;45:1732–1740.
3 Koffler D, Agnello V, Winchester R, Kunkel HG: The occurrence of single-stranded DNA in the serum of patients with systemic lupus erythematosus and other diseases. J Clin Invest 1973;52: 198–204.
4 Steinman CR: Free DNA in serum and plasma from normal adults. J Clin Invest 1975;56:512–515.
5 Rumore M, Steinman CR: Endogenous circulating DNA in systemic lupus erythematosus, occurrence as multimeric complexes bound to histone. J Clin Invest 1990;86:69–74.
6 Li ZJ, Steinman CR: Plasma DNA in systemic lupus erythematosus: Characterization of cloned base sequences. Arthritis Rheum 1989;32:726–733.
7 Leon SA, Shapiro B, Sklaroff DM, Yaros MJ: Free DNA in the serum of cancer patients and the effect of therapy. Cancer Res 1977;37:646–650.

8 Stroun M, Anker P, Lyautey J, Lederrey C, Maurice PA: Isolation and characterization of DNA from the plasma of cancer patients. Eur J Cancer Clin Oncol 1987;23:707–712.

9 Maebo A: Plasma DNA level as a tumor marker in primary lung cancer. Jpn J Thoracic Dis 1990; 28:1085–1091.

10 Fournié GJ, Courtin JP, Laval F: Plasma DNA as a marker of cancerous cell death. Investigation in patients suffering from lung cancer and in nude mice bearing human tumour. Cancer Lett 1995;2: 221–227.

11 Stroun M, Anker P, Maurice P, Lyautey J, Lederrey C, Beljanski M: Neoplastic characteristics of the DNA found in the plasma of cancer patients. Oncology 1989;46:318–322.

12 Anker P, Mulcahy H, Chen XQ, Stroun M: Detection of circulating tumour DNA in blood (plasma/ serum) of cancer patients. Cancer Metastasis Rev 1999;18:65–73.

13 Stroun M, Maurice P, Vasioukhin V, Lyautey J, Lederrey C, Lefort F, Rossier A, Chen XQ, Anker P: The origin and mechanism of circulating DNA. Ann NY Acad Sci 2000;906:161–168.

14 Vasyoukhin V, Anker P, Maurice P, Lyautey J, Lederrey C, Stroun M: Point mutations of the N-ras in the blood plasma DNA of patients with myelodysplastic syndrome or acute myelogenous leukaemia. Br J Haematol 1994;86:774–779.

15 Ryan BM, McManus RO, Daly JS, Keeling PW, Weir DG, Lefort F, Kelleher D: Serum mutant K-ras in the colorectal adenoma-to-carcinoma sequence. Ann NY Acad Sci 2000;906:29–30.

16 Mulcahy H, Anker P, Lyautey J, Chen Xq, Lederrey C, Farthing M, Stroun M: K-ras gene mutations in the plasma of pancreatic patients. Clin Cancer Res 1998;4:271–275.

17 Yamada T, Nakamori S, Ohzato H, Oshima S, Aoki T, Higaki N, Sugimoto K, Agaki K, Fujiwara Y, Nishisho I, Sakon M, Gotoh M, Monder M: Circulating DNA K-ras mutation in pancreatic adenocarcinoma. Clin Cancer Res 1998;4:1527–1532.

18 Frickhofen N, Müller E, Sandherr M, Binder T, Bangerter M, Wiest C, Enz M, Heimper H: Rearranged Ig heavy chain DNA is detectable in cell free blood samples of patients with B-cell neoplasia. Blood 1997;90:4953–4960.

19 Chen XQ, Stroun M, Magnenat JL, Nicod LP, Kurt AM, Lyautey J, Lederrey C, Anker P: Microsatellite alterations in plasma DNA of small cell lung cancer patients. Nature Med 1996;2:1033–1035.

20 Nawroz H, Koch W, Anker P, Stroun M, Sidransky D: Microsatellite alterations in plasma DNA of head and neck cancer patients. Nature Med 1996;2:1035–1037.

21 Sanchez-Cespedes M, Monzo M, Rosell R, Pifarré A, Calvo R, Lopez-Cabrerizo MP, Astudillo J: Detection of chromosome 3p alterations in serum DNA of non-small-cell lung cancer patients. Ann Oncol 1998;9:113–116.

22 Goessl C, Heicappel R, Munker R, Anker P, Stroun M, Krause H, Muller M, Miller K: Microsatellite analysis of plasma DNA from patients with clear cell renal carcinoma. Cancer Res 1998;58: 4728–4732.

23 Utting M, Müller G, Werner W, Schubert J, Junker K: Detection of tumor genetic alterations of bladder carcinomas in body fluids depends on sample treatment before DNA isolation. Ann NY Acad Sci 2000;906:67–72.

24 Silva J, Dominguez G, Garcia JM, Gonzalez R, Villanueva MJ, Navarro F, Provencio M, Martin SST, Espana P, Bonilla F: Presence of tumor DNA in plasma of breast cancer patients: Clinicopathological correlations. Cancer Res 1999;59:3251–3256.

25 Kölble K, Ullrich OM, Pidde H, Barthel B, Diermann J, Rudolph B, Dietel M, Schlag PM, Scherneck S: Microsatellite alterations in serum DNA of patients with colorectal cancer. Lab Invest 1999; 79:1145–1150.

26 Hickey KP, Boyle DP, Jepps HM, Andrew AC, Buxton EJ, Burns PA: Molecular detection of tumour DNA in serum and peritoneal fluid from ovarian cancer patients. Br J Cancer 1999;80: 1803–1808.

27 Fujiwara Y, Chi DD, Wang H, Keleman P, Morton DL, Turner R, Hoon DS: Plasma DNA microsatellites as tumor-specific markers and indicators of tumor progression in melanoma patients. Cancer Res 1999;59:1567–1571.

28 Mutirangura A, Pornthanakasem W, Theamboonlers A, Sriuranpong V, Lertsanguansinchi P, Yenrudi S, Voravud N, Supiyaphun P, Poovorawan Y: Epstein-Barr viral DNA in serum of patients with nasopharyngeal carcinoma. Clin Cancer Res 1998;4:665–669.

29 Esteller M, Sanchez-Cespedes EM, Rosell M, Sidransky D, Baylin SB, Herman JG: Detection of aberrant promoter hypermethylation of tumor suppressor genes in serum DNA from non-small cell lung cancer patients. Cancer Res 1999;59:67–70.

30 Wong IH, Lo YM, Zhang J, Liew CT, Ng MH, Wong N, Lai PB, Lau WY, Hjelm NM, Johnson PJ: Detection of aberrant p16 methylation in the plasma and serum of liver cancer patients. Cancer Res 2000;59:71–73.

31 Eisenberger CF, Schoenberg M, Enger C, Hortopan S, Shah S, Chow NH, Marshall FF, Sidransky D: Diagnosis of renal cancer by molecular urinalysis. J Natl Cancer Inst 1999;91:2028–2032.

Philippe Anker, PhD, Department of Plant Biochemistry and Physiology
University of Geneva, Pavillon des Isotopes, 20 bvd d'Yvoy, CH–1211 Geneva (Switzerland)
Tel. +41 22 827 23 16, Fax +41 22 781 51 93, E-Mail Philippe.Anker@bota.unige.ch

Hahn S, Holzgreve W (eds): Fetal Cells and Fetal DNA in Maternal Blood. New Developments for a New Millennium. 11th Fetal Cell Workshop, Basel, 2000. Basel, Karger, 2001, pp 8–15

Experience with Single Cell PCR in Preimplantation Genetic Diagnosis: How to Avoid Pitfalls

S. Rechitsky, O. Verlinsky, C. Strom, V. Ivakhnenko, J. Nefedova, M. Rechitsky, T. Kouliev, T. Amet, J. Pomeranetz, Y. Verlinsky, A. Kuliev

Reproductive Genetics Institute, Chicago, Ill., USA

Introduction

Preimplantation genetic diagnosis (PGD) involves testing single cells biopsied from oocytes or embryos. Accordingly, the accuracy of PGD depends on the controlling of well-known pitfalls of single-cell PCR analysis. These include contamination by extraneous DNA, the preferential amplification phenomenon and allele-drop out (ADO). It has been demonstrated that ADO rates in single-cell PCR may vary with different lysis procedures, cell types and loci analyzed [1, 2]. Therefore, reliable methods are needed to detect potential ADO or preferential amplification in order to avoid misdiagnosis in single-cell PCR. We have previously reported preliminary experience in the detection of ADO by a sequential analysis of oocytes by the first and second PB (PB1 and PB2), and by simultaneous amplification of mutant genes with linked polymorphic markers [2, 3].

In the present chapter, we review our current experience emphasizing the approaches for detection of ADO to achieve a high accuracy of PCR analysis acceptable for clinical application in PGD.

Material and Methods

A list of the genetic disorders tested using the single-cell PCR system is presented in table 1. Materials for testing were PB1 and PB2, removed from oocytes, single blastomeres biopsied from 8-cell embryos, or single fibroblasts isolated by macromanipulation from skin

Table 1. List of conditions analyzed by single-cell PCR for preimplantation genetic diagnosis

Cystic fibrosis
Thalassemia
Sickle cell disease
Tay-Sachs disease
Phenylketonuria
Hemophilia A and B
Gaucher disease
Long-chain 3-hydroxyacyl-CoA dehydrogenase deficiency
Epidermolis bullosa
Retinitis pigmentosa
Ornithine-transcarbamylase deficiency
Alport disease
α_1-Antitrypsin deficiency
Multiple epiphyseal dysplasia
X-linked hydrocephalus
Myotonic dystrophy

biopsies of heterozygous carriers. The biopsied single cells were placed directly into a lysis solution, consisting of 0.5 μl 10 × PCR buffer, 0.5 μl 1% Tween 20, 0.5 μl 1% Triton X-100, 3.5 μl H_2O and 0.05 μl protienase K (20 mg/ml in a 0.5-ml PCR tube). After spinning down at a low speed in a microfuge for a few seconds, the samples were covered with 1 drop of mineral oil and incubated at 45 °C for 15 min in a thermal cycler. Proteinase K was then inactivated at 96 °C for 20 min, which is also the beginning of the hot start of the first-round PCR. In case samples had to be shipped over, they were covered with 1 drop of mineral oil and frozen in vertical position. Overall, more than 2,000 single-cell PCR reactions were performed for PB1, PB2 and single blastomeres, using multiplex nested PCR analysis. Lower stringency and longer annealing time were used in the first-round PCR, with the introduction of the mixture of all outside primers for both mutant genes and polymorphic markers. Following the first-round PCR, separate aliquots were amplified in the second-round PCR with specific inside primers for each site, using a higher stringency. Such a dual or multiple amplification reaction allowed detection of most of the ADO cases. To eliminate false priming to possible pseudogenes, as in PGD of Gaucher disease or long-chain acyl-CoA deficiency (LCHAD), the first-round primers were designed to anneal to the regions of nonidentity with pseudogenes [2]. Primer sequences, reaction conditions and details of nested PCR reactions for PGD of the disorders listed in table 1 have been described elsewhere [1–5]. In addition to short-tandem repeats (STRs) linked to the genes studied, STRs located on other chromosomes were also studied to test for a possible contamination by extraneous DNA, and identification of the origin of individual embryos in the established pregnancies. A list of STRs, their sequences and PCR conditions for their analysis were also reported previously [1–5].

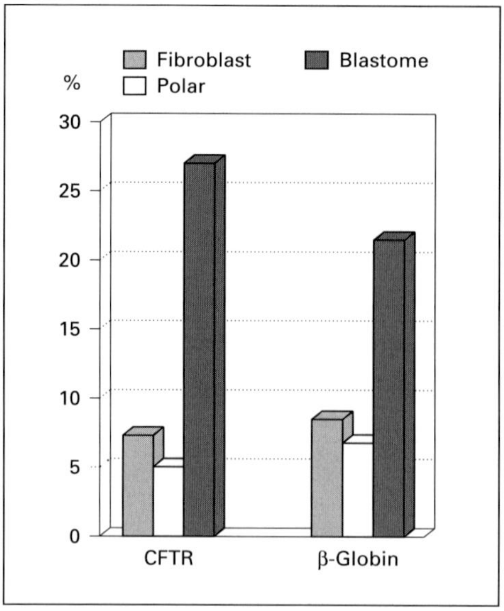

Fig. 1. ADO rate in different types of single cells.

In some cases fluorescent PCR (F-PCR) was also performed for a direct fragment size analysis of the PCR product [6]. Because F-PCR is known to be much more sensitive than the standard gel detection, a fewer PCR cycles were run, with multiplex reaction performed and analyzed in one tube, using different color labels on each primer. F-PCR was particularly useful also for direct sequencing of the PCR product in the detection of point mutations and for distinguishing preferential amplification from ADO.

To investigate the accuracy of PGD, the embryos resulting from the oocytes predicted to contain a mutant maternal allele, or those with insufficient polymorphic marker information, were exposed to confirmatory PCR analysis. In the established pregnancies the results of PGD were confirmed by prenatal diagnosis.

Results and Discussion

Our experience shows that a single-cell DNA analysis is more accurate and reliable in PB and fibroblasts than in single blastomeres (fig. 1), ADO being at least twice less frequent in PB1 than in blastomeres. In addition, a considerable proportion of ADO is also detectable by sequential analysis of PB1 and PB2 [3]. Data on sequential PCR analysis of 16 alleles in PB1 and PB2 obtained from 794 oocytes showed that more than half of all ADOs are detected simply by sequential analysis of PB1 and PB2 [2, 3]. This allows avoiding misdiagnosis due to ADO

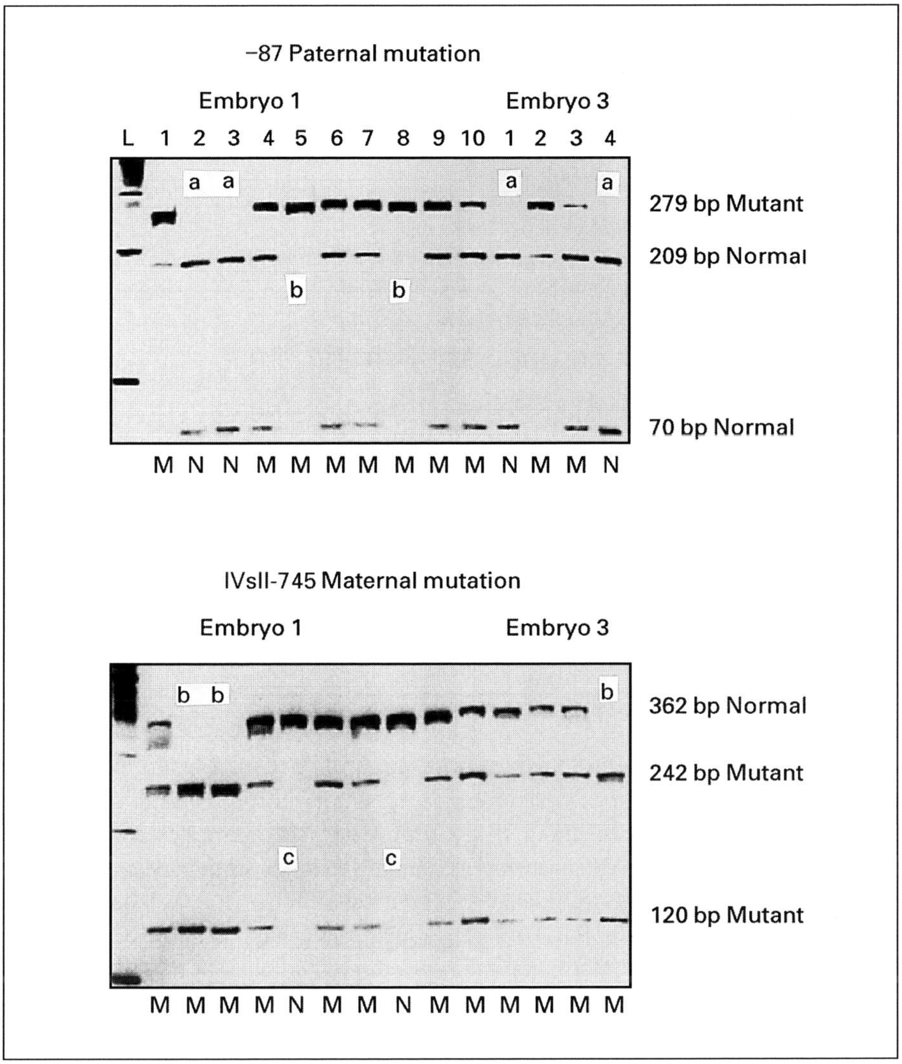

Fig. 2. ADO: possible cause of misdiagnosis in compound heterozygous blastomeres. [a] ADO of −87 allele. [b] ADO of normal allele. [c] ADO of IVS II 745 allele.

even in those cases when no informative polymorphic markers are available, as in cases of PGD in some isolated populations and ethnic groups [4, 5]. As can be seen from figure 2, a high rate of ADO in blastomeres may lead to an obvious misdiagnosis in compound heterozygous embryos. In fact, most of the misdiagnoses described at the initial stage of application of PGD for single-gene disorders

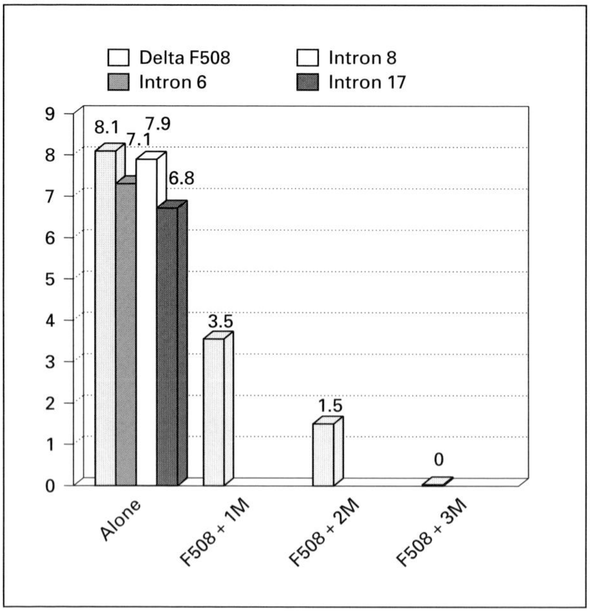

Fig. 3. Power of linked markers in ADO detection.

were in cases of blastomere biopsy from apparently compound heterozygous embryos [7–10]. However, if two or more linked polymorphic markers are available for testing of the gene under study, almost all ADOs are detected, making DNA testing in PB or single blastomeres a highly reliable procedure (fig. 3).

The application of F-PCR also makes it possible to improve detection of potential misdiagnoses in PGD of single-gene disorders. Because preferential amplification, similar to ADO, may be represented by a lack of one of the alleles in a single heterozygous cell, analyzed by a conventional PCR analysis, some of the ADOs may actually be due to preferential amplification. We introduced F-PCR to distinguish ADO from preferential amplification in single heterozygous cells for CF Delta F-508 mutation. Testing 148 single fibroblasts by both conventional and F-PCR showed that preferential amplification has a minor contribution to misdiagnosis (1% for Delta F-508 mutation and 0.9% for intron 6 polymorphism; fig. 4). Simultaneous amplification of single cells for CF Delta F-508 mutation together with one linked polymorphic marker reduced the ADO rate by more than half, regardless of the use of conventional or F-PCR. With the additional second marker in multiplex PCR the ADO rate was further reduced by half, being completely absent if 3 markers were simultaneously amplified, as shown above in figure 3.

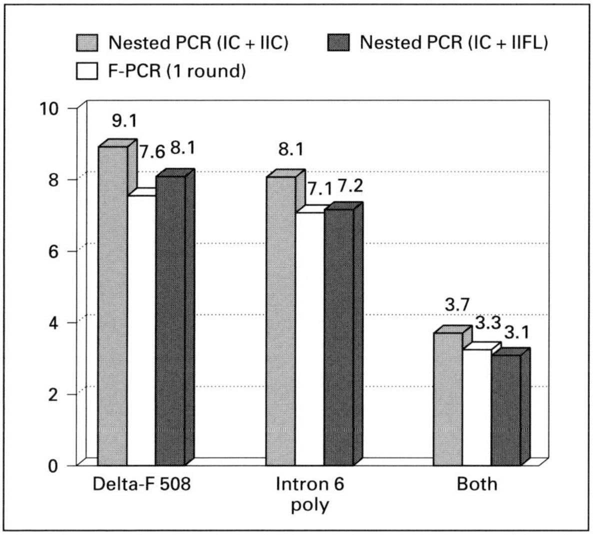

Fig. 4. Comparison of ADO rates following fluorescent or conventional PCR.

Table 2. Results of two-step polar body analysis in predicting the genotype of oocytes

Loci	Total oocytes amplified	Oocytes with IPB data		Oocytes with IIPB data	Oocytes with IPB and IIPB data		Resulting embryos	
		hetero-zygote	homo-zygote		total	ADO	studied	confirmed
Mutations (n = 16)	794	527	266	531	497	32	251	245 (98%)
STRs (n = 10)	437	298	139	324	304	22	112	111 (99%)
Total mutations and STRs	1,231	825	405	1,006	801	54	363	356 (98%)

By the present time we performed PGD in 69 cycles for couples at high risk of having children with single-gene disorders, providing the possibility to preselect and transfer a sufficient number of mutation-free oocytes in almost all cycles. Of 794 oocytes with DNA results, 527 (66.4%) had heterozygous PB1, i.e. with both normal and mutant genes amplified, therefore ideal for further testing, although their potential transfer depended entirely on the identification of mutant gene in the sequential analysis of PB2 (table 2). Thus, a priority in preselection of embryos for transfer were given to the embryos resulting from the oocytes with heterozygous PB1 because in the absence of DNA contamination this indicates the absence of ADO of either the normal or mutant allele. Although most of the transferred embryos were preselected using this particular strategy, some preselected embryos still originated from homozygous normal oocytes, inferred from homozygous mutant status of PB1 and hemizygous normal status of PB2. Because the number of cross-over events in meiosis resulting in the heterozygous PB1 depends strongly on the distance between the gene studied and the centromere, no heterozygous oocytes were observed in Gaucher's disease gene testing, mapped close to the centromere of chromosome 1. These embryos were accepted for transfer only if ADO could have been excluded using linked polymorphic marker analysis. Otherwise, such embryos were excluded from transfer and exposed to follow-up confirmation analysis of the resulting embryos. In fact, at least 8 oocytes initially inferred as normal from the finding of a homozygous mutant PB1 and a hemizygous normal PB2 appeared to be affected following the analysis of linked markers, indicating ADO of a normal allele in an apparently heterozygous PB1 in these cases.

This approach made it possible to detect ADO in 32 of 794 oocytes tested for mutations, which could have contributed to misdiagnosis should they be used for transfer. The follow-up analysis of the embryos excluded from transfer either because they were affected, or there was insufficient information to preselect them for transfer, has provided valuable data for evaluating the proportion of undetected ADO. Overall, 61 (6.9%) ADOs were observed, which include 54 detected and 7 undetected ones, suggesting that actually 886 oocytes were heterozygous, in contrast to 827 initially identified by PB1 analysis for mutations and STR (table 2). As seen from table 2, the genotype of 6 embryos appeared to be different from the predicted by mutation analysis, and 1 by STR, due to ADO in PB1, and they were diagnosed as homozygous instead of their actual heterozygous status. The data indicate 98% accuracy, which is quite acceptable for the clinical use of PGD for single-gene disorders (table 2).

The work has presently resulted in preselection and transfer of mutation-free embryos in almost all cycles, with 25 established pregnancies, resulting in the birth of 20 healthy children.

PGD has also been provided for couples with one homozygous affected partner, including thalassemia and PKU, resulting in unaffected pregnancies and the birth of healthy children in the latter case (unpublished data). Further optimization of the method will allow completely avoiding the risk for misdiagnosis due to ADO and other limitations of single-cell PCR analysis in PGD for single-gene disorders. With these improvements in the accuracy and reliability of the single-cell DNA testing, its application to PGD is presently extended for those conditions that have never been considered for PGD, such as cancer predisposition or HLA matching. As in PGD for chromosomal disorders, the present status of PGD for single-gene disorders and those conditions determined by genetic predisposition makes PGD not only an alternative but also an important complement to prenatal diagnosis, providing an option for more ethically acceptable approaches for prevention of genetic diseases.

References

1 Rechitsky S, Strom C, Verlinsky O, Amet T, Ivakhnenko V, Kukharenko V, Kuliev A, Verlinsky Y: Allele drop out in polar bodies and blastomeres. J Assist Reprod Genet 1998;15:253–257.

2 Rechitsky S, Strom C, Verlinsky O, Amet T, Ivakhnenko V, Kukharenko V, Kuliev A, Verlinsky Y: Accuracy of preimplantation diagnosis of single-gene disorders by polar body analysis of oocytes, J Assist Reprod Genet 1999;16:169–175.

3 Verlinsky Y, Rechitsky S, Verlinsky O, Ivakhnenko V, Lifchez A, Kaplan B, Moise J, Valle J, Borkowski A, Nefedova J, Goltsman E, Strom C, Kuliev A: Prepregnancy testing for single-gene disorders by polar body analysis. Genet Test 1999;3:185–190.

4 Kuliev A, Rechitsky S, Verlinsky O, Ivakhnenko V, Evsikov G, Wolf G, Angastiniotis M, Georghiou D, Kukharenko V, Strom C, Verlinsky: Preimplantation diagnosis of thalassemias. J Assist Reprod Genet 1998;15:219–225.

5 Kuliev A, Rechitsky S, Verlinsky O, Ivakhnenko V, Cieslak J, Evsikov S, Wolf G, Angastiniotis M, Kalakoutis G, Strom C, Verlinsky Y: Birth of healthy children after preimplantation diagnosis of thalassemias. J Assist Reprod Genet 1999;16:207–211.

6 Findlay I, Quirke P: Fluorescent polymerase chain reaction: Part I. A new method allowing genetic diagnosis and DNA fingerprinting of single cells. Hum Reprod Update 1996;2:137–152.

7 Verlinsky Y, Kuliev AM (eds): Preimplantation Diagnosis of Genetic Diseases: A New Technique in Assisted Reproduction New York, Wiley-Liss, 1993.

8 Harper J: Preimplantation diagnosis of inherited disease by embryo biopsy. An update of the world figures. J Assist Reprod Genet 1996;13:90–95.

9 Handyside AH, Delhanty JDA: Cleavage stage biopsy of human embryos and diagnosis of X-linked recessive disease; in Edwards RG (ed): Preconception and Preimplantation Diagnosis of Human Genetic Disease. Cambridge, Cambridge University Press, 1993, p 239.

10 Grifo JA, Tang YX, Munne S, et al: Healthy deliveries from biopsied human embryos. Hum Reprod 1994;9:912–916.

Anver Kuliev, 2825 North Halsted, Chicago, IL 60657 (USA)
Tel. +1 773 296 7095/+1 773 472 4900, ext. 2163, Fax +1 773 935 8087/+1 773 871 5221
E-Mail Anverkuliev@hotmail.com

Hahn S, Holzgreve W (eds): Fetal Cells and Fetal DNA in Maternal Blood. New Developments for a New Millennium. 11th Fetal Cell Workshop, Basel, 2000. Basel, Karger, 2001, pp 16–20

Laser-Mediated Micromanipulation Systems: An Important Tool for the Analysis of Single Cells

Xiao Yan Zhong, Wolfgang Holzgreve, Sinuhe Hahn

Laboratory for Prenatal Medicine, Department of Obstetrics and Gynecology, University of Basel, Switzerland

Introduction

The molecular biological examination of small clinical samples has been hampered by the presence of undesired contaminating background cells which obscure the analysis of target cells of interest [1]. Recent advances in molecular biological techniques, such as the polymerase chain reaction (PCR), permits the analysis of a few or even single cells. In this manner, fetal cells isolated from maternal blood can be used as a noninvasive source of fetal genetic material for prenatal diagnosis [2]. A caveat of this and other similar approaches is that the analysis of specific target cell DNA or RNA is currently hindered by the inability to isolate the target cells of interest from the tissue sample biopsied. In this manner, even though single-cell PCR is a powerful method for determining the genetic properties of individual cells [3], the successful analysis of rare fetal cells in the maternal circulation depends on the ability to reliably separate these from any residual contaminating maternal cells.

One way in which this has been attempted is by micromanipulation, whereby the cell was simply scraped with a finely drawn glass needle from the glass slide [4–6]. This technique was found to be unsatisfactory since it was not always easy to cleanly remove the cell from the slide. Furthermore, frequently the cells would shear and it would not be possible to determine whether the entire cell had been manipulated [7]. The development of laser micromanipulation has revolutionized microdissection [8, 9]. The aim of this brief review is to familiarize investigators in prenatal medicine with these new micromanipulation systems.

Laser Capture Microdissection System

Laser capture microdissection (LCM) was developed by Liotta and colleagues [8, 9] at the National Institutes of Health, as a new method for the separation of groups of cells or even single cells from tissue sections or cytological smears, thereby resulting in pure cell preparations for DNA, RNA, or protein analysis [9, 10]. In Europe this system is marketed by MWG Biotech. The LCM system combines a standard laboratory microscope with a 980-nm continuous IR-diode laser. For the manipulation of the target cell, an optical cap, the size of a standard PCR vessel, which is coated with a 5-μm transfer film is positioned directly onto the tissue surface for cytological preparation. Once the target cell has been selected, a focused laser pulse is applied to it which selectively activates the transfer film directly above the target cell. This pulse of energy then covalently links the cell(s) in the activated region directly to the transfer film. The process can then be repeated to collect more desired target cells or the optical cap can be placed directly on a standard PCR tube permitting any of a number of molecular examinations. Upon the application of a lysis buffer, the cellular material detaches from the film and is ready for subsequent molecular analysis. The whole procedure takes only a few minutes and presents a very soft and gentle isolation method. Since the exact and native morphologies of both the captured cells and the surrounding tissues are preserved, therefore even enzymatic assays are possible.

Typical application areas for LCM are cancer research, gene expression studies, research programs on multiple sclerosis, arteriosclerosis or Alzheimer's disease, loss of heterozygosity and microsatellite instability studies and more recently, proteomics (2D protein gels, Western blotting). By using the LCM system, it has been possible to transfer cells from paraffin-embedded and frozen tissue samples, as well as histologically stained or immunohistochemically labelled and stained slides and cytological smears.

In our group, we have tested this system for isolation and molecular characterization of single erythroblasts enriched from maternal blood samples.

In our hands we found the system to be easy to use; however, it was difficult to reliably isolate single cells. This is due in part to the physical contact of the membrane in the optical cap with the entire cytological preparation, whereby cells can be transferred nonspecifically. In addition, the large volume required for efficient lysis of the single cell was found to hinder the subsequent PCR considerably. A slight deficit of this system is that it is currently not compatible with many of the cytological stains used for the analysis of fetal erythroblasts, including the May-Grünwald-Giemsa stain.

Laser Pressure Catapulting

The laser pressure catapulting (LPC) system (Robot-MicroBeam), marketed by PALM Microlaser Technology, consists of a 337-nm high-quality pulsed nitrogen laser coupled to the epifluorescent path of an inverted microscope which is focused through a high numerical objective to yield a spot size of < 2 μm in diameter. The high photonic pressure force of the focused laser beam ejects the selected sample from the object plane and catapults it directly into the cap of a low-cost routine PCR tube. Subsequently, the catapulted specimen is dissolved by a lysis buffer and the cellular contents are centrifuged into the bottom of the tube.

Using the LPC system it is possible to transfer single or groups of cells from frozen or fixed tissue sections, cytospins, cell smears, cultured cells, chromosomal spreads and DNA filaments to procure cell areas, single cells, chromosomes, chromatids or chromosomal parts [11–13].

In our group, we tested this system for analysis of single erythroblasts enriched from maternal blood samples. In our experience, the micromanipulation of single cells from cytospin preparations by the LPC system was shown to be very easy and quick. An advantage of this system is that all common cell or tissue preparation techniques can be used. For our work we have tried both the transfer of single cells from normal glass slides as well as from cell preparations which have been made on special thin plastic film fixed on the glass slides. This latter feature permits the easier cutting out of single cells from dense preparations and also permits a better retention of the cell shape upon laser catapulting.

The advantages of the LPC 'non-contact' technique is that it is not as prone to spurious contaminants as the LCM system, which relies on direct contact of a polyester film with the cytological preparation. Furthermore, the extremely high photon density in the narrow focus of the pulsed nitrogen laser allows the system to cut or ablate biological structures with relative ease. Since this laser cutting is a locally restricted ablative form of a photo-decomposition process which occurs without heating, it is thus possible to perform microdissection of large biological molecules (chromosomes) or the micropreparation of single cells and subcellular particles without damaging adjacent regions or impairing cell viability.

Other Laser Micromanipulating systems

Several other laser-mediated systems have recently emerged on the market, amongst them the UV CUT laser microdissection system marketed by SL Microtest in Germany. This system is also a highly sophisticated microscopy tool with which one is able to isolate small areas of tissue or single cells out of a histological section. Akin to the other described systems, the UV CUT system is based on high-quality components including an inverted microscope, a motorized xy-stage, 3-chip CCD camera and maintenance-free laser. In a similar manner to the LPC system, the incision path is marked with a drawing tool on the monitor by the aid of a computer mouse, which is then automatically cut out once correctly assigned. A unique feature of this system is the 'single-step collection' method, whereby a specific adhesive film in the lid of a reaction tube picks up the dissected cells and allows the recovery of several dissected cells or areas, even from different slides, onto one reaction tube. Similar to the other systems described, it is also possible here to examine frozen or fixed tissue sections, cell smears and cytospins.

The Leitz/Leica Company has also recently released a laser microdissection system, which we have not been able to inspect more closely. This system uses a pulsed UV laser coupled to a Leica automated laboratory microscope for observation and dissection. The main difference of this system is that once the target cells are cut out by the action of the UV laser, they are transferred by gravity into the PCR tube which is located below the dissection platform. A special lens system enables the operator to visually ensure that the target cells have indeed been transferred to the PCR tube.

Conclusion

Single-cell PCR has established itself as a useful method in modern medicine and biology, for instance in cancer research, prenatal diagnosis and genetic analysis. Numerous efforts have been made to selectively retrieve small groups of defined cells from heterogeneous mixtures or tissue samples. Previously these efforts have made use of mechanical methods using microcapillaries or microneedles to micromanipulate single cells from stained tissue sections or cytological preparations for subsequent PCR analysis. These methods are not completely satisfactory as the preparation of the needles and the contamination-free micromanipulation of the cells is very time consuming. In addition it was near to impossible to accurately obtain single cells from paraffin-embedded or other archival material using these mechanical techniques. Recently several companies have developed laser microdissection systems to provide a reliable method for the isolation of pure populations of single cells from specific microscopic regions of tissue sections and cytospins, in a single-step manner which takes place under direct visualization [9–11]. Our aim has been to briefly elude to the strengths and weaknesses of these systems for the efficient retrieval of single fetal cells from maternal blood. Our own experience has shown that laser microdissection systems offer several advantages over current cell micromanipulation approaches. In comparison with the micromanipulation using microcapillaries or microneedles, they are simple to use and much faster, and also certainly more precise. By using either of the discussed systems it is possible to directly punch out or cut out single cells from a cluster of cells, a feature which is almost impossible to duplicate using conventional mechanical micromanipulation. In addition, the use of sterile, disposable transfer film-coated optical caps, or the direct catapulting of the cell into the PCR cap, considerably reduces the risk of contamination. Therefore, these systems have the potential to fulfill the critical needs of basic researchers and clinicians who need to perform single-cell manipulation on a routine basis.

Acknowledgements

This work was supported by the Swiss National Science Foundation (3200-059275.99/1), and the National Institutes of Health (contract number N01-HD-4-3202).

References

1 Krismann M, Todt B, Schröder J, Gareis D, Müller KM, Seeber S, Schütte J: Low specificity of cytokeratin 19 reverse transcriptase-polymerase chain reaction analyses for detection of hematogenous lung cancer dissemination. J Clin Oncol 1995;13:2769–2775.

2 Hahn S, Sant R, Holzgreve W: Fetal cells in maternal blood: Current and future perspectives. Mol Hum Reprod 1998;4:515–521.

3 Garvin AM, Holzgreve W, Hahn S: Highly accurate analysis of heterozygous loci by single cell PCR. Nucleic Acids Res 1998;26:3468–3472.

4 Takabayashi H, Kuwabara S, Ukita T, Ikawa K, Yamafuji K, Igarashi T: Development of non-invasive fetal DNA diagnosis from maternal blood. Prenat Diagn 1995;15:74–77.

5 Hahn S, Zhong XY, Troeger C, Burgemeister R, Gloning K, Holzgreve W: Current applications of single-cell PCR cell. Cell Mol Life Sci 2000;57:96–105.

6 Hahn S, Zhong XY, Bürk MR, Troeger C, Holzgreve W: Multiplex and realtime quantitative PCR on fetal DNA in maternal plasma: A comparison with fetal calls isolated from maternal blood. Ann NY Acad Sci 2000;906:148–152.

7 Goring JJ, Lamb RF: Practical histological microdissection for PCR analysis. J Pathol 1996;179: 121–124.

8 Peterson LA, Brown MR, Carlisle AJ, Kohn EC, Liotta LA, Emmert-Buck MR, Krizman DB: An improved method for construction of directionally cloned cDNA libraries from microdissected cells. Cancer Res 1998;58:5326–5328.

9 Emmert-Buck MR, Bonner RF, Smith PD, Chuaqui RF, Zhuang Z, Goldstein SR, Weiss RA, Liotta LA: Laser capture microdissection. Science. 1996;274:998–1001.

10 Luo L, Salunga RC, Guo HQ, Bittner A, Joy KC, Galindo JE, Xiao HN, Rogers KE, Wan JS, Jackson MR, Erlander MG: Gene expression profiles of laser-captured adjacent neuronal subtypes. Nat Med 1999;5:117–122.

11 Schütze K, Pösl H, Lahr G: Laser micromanipulation systems as universal tools in molecular biology and medicine. Cell Mol Biol (Noisy-le-Grand) 1998;44:735–746.

12 Bernsen MR, Dijkman HBPM, Vries E, Figdor CG, Ruiter DJ, Adema GJ, Muljin GNP: Identification of multiple mRNA and DNA sequences from small tissue samples isolated by laser assisted microdissection. Lab Invest 1998;78:1267–1273.

13 Fink L, Seeger W, Ermert L, Hänze J, Stahl U, Grimminger F, Kummer W, Bohle RM: Real-time quantitative RT-PCR after laser-assisted cell picking. Nature Med 1998;4:1329–1333.

Dr. Sinuhe Hahn, Laboratory for Prenatal Medicine, Department of Obstetrics and Gynecology
University of Basel, Schanzenstrasse 46, CH–4031 Basel (Switzerland)
Tel. +41 61 325 9224/9595, Fax +41 61 325 9399, E-Mail shahn@uhbs.ch

Hahn S, Holzgreve W (eds): Fetal Cells and Fetal DNA in Maternal Blood. New Developments for a New Millennium. 11th Fetal Cell Workshop, Basel, 2000. Basel, Karger, 2001, pp 21–27

Why Do We Need Noninvasive Prenatal Diagnostic Procedures?

Daniel V. Surbek, Wolfgang Holzgreve

Department of Obstetrics and Gynecology, University of Basel, Switzerland

Introduction

Today, prenatal diagnosis of chromosomal anomalies and single gene disorders is based on invasive procedures. Different techniques have been developed since the first prenatal diagnosis of chromosomal anomalies by amniocentesis in the late 1960s [1].These procedures basically consist of sampling of tissue or cells from the placenta, fetus or amniotic fluid by ultrasound-guided insertion of a needle into the pregnant uterus. Although current techniques for invasive diagnosis are generally regarded as relatively safe, if performed with appropriate technical equipment by an experienced operator, these procedures carry a risk for loss of pregnancy in the magnitude of around 1%. On the other hand, the expected rate of major chromosomal anomalies in live births of a given population is 0.65%, including trisomy 21 with a frequency of 0.12% [2]. For this reason, invasive testing should be confined to a selected population at high risk for chromosomal anomalies, in order to reduce the total number of chromosomally normal, healthy fetuses lost due to the procedure.

On this background, strategies have been developed to estimate the risk for chromosomal anomalies in individual pregnancies using noninvasive testing which does not carry a risk to the pregnancy itself. Initially, maternal age has been used to estimate the risk of chromosomal anomalies based on the fact that the risk increases with age. A cutoff level of around 35 years (when the risk of Down's syndrome is around 1 in 280) has been used 'historically'. However, if all women above 35 years undergo an invasive procedure, which corresponds to 5–7% of all pregnant women, the detection rate for fetuses with Down's syndrome is only between 30 and 40%. In the 1980s and 1990s, the influence of chromosomally abnormal fetuses on certain hormones and proteins circulating in maternal blood was detected. In addition, ultrasound features of the fetus such as increased nuchal translucency have been described which proved to be associated with aneuploidy. Based on this new knowledge, strategies have been developed in search of a more precise and individual assessment of the risk of chromosomal anomalies, as compared to the sole use of the age of the mother. With these strategies, detection rates of 60–70% with second-trimester maternal serum biochemistry and 80–90% using first-trimester ultrasound (nuchal translucency measurement) and biochemistry, each with a 5% invasive procedure rate, are being achieved.

This now means that it is possible today to target invasive procedures much better on the population at risk. Nevertheless, thousands of pregnancies with healthy fetuses are still lost every year in the world due to invasive procedures. The challenge to find ways to diagnose chromosomal anomalies without putting the fetus at risk remains.

A variety of different ultrasound-guided sampling techniques to obtain cells for fetal karyotyping are available. They include conventional amniocentesis, chorionic villus sampling (CVS), early amniocentesis, placental biopsy (late CVS), and fetal blood sampling. These methods differ regarding the gestational age at which they are performed, operator experience necessary, time needed until results are obtained, and risk to the fetus. For example, CVS is being performed earlier and provides faster results than conventional amniocentesis. This means that a procedure must be selected on the basis of the individual situation, taking into account the above-mentioned factors as well as the a priori risk for a chromosomal anomaly (or a single gene disorder) in the fetus. The risk of the procedure itself can be deduced from large-scale multicenter trials.

The aim of this chapter is to give an overview of the current techniques used today for invasive fetal karyotyping or diagnosis of single-gene disorders, with a special view to the procedure-related risks.

Techniques of Invasive Procedures

Amniocentesis

Over 30 years ago the first reports on the detection of chromosomal anomalies in the fetus by karyotyping of amniotic fluid cells appeared in the literature [1, 3]. Initially, amniotic fluid was obtained 'blindly'. In the late 1970s and early 1980s, ultrasound was introduced, initially as a tool to detect placental free amniotic fluid pockets to determine the optimal entry site. Later, continuous ultrasound for needle guidance was developed and is still being used. The procedure is usually performed between 15 and 18 weeks of gestation. The needle is inserted under continuous ultrasound visualization of the needle tip, either in a 'free-hand' technique, or with a needle-guide which is attached to the ultrasound transducer. Although insertion through the placenta is avoided whenever possible, comparing studies do not suggest an increased risk of fetal loss with the transplacental procedure [4].

Many studies have assessed the procedure-related risk of amniocentesis. The best evidence on the magnitude of risk comes from a Danish randomized trial where the fetal loss rate was 1% higher as compared to no procedure [5]. Long-term, population-based follow-up studies of newborns after amniocentesis have not shown any abnormalities, except an increased risk of hemolytic disease due to isoimmunization [6]. A collaborative study in the US revealed that there is a clear correlation between the number of needle insertions and the risk of fetal loss [7]. Other studies assessed whether there is an incidence of orthopedic problems in the newborns but did not find an association with amniocentesis.

Chorionic Villus Sampling

CVS has been developed as an alternative to amniocentesis. It is performed significantly earlier in pregnancy, between 11 and 14 weeks, and has the advantage that more cells are obtained and are available for analysis. Rapid karyotyping by conventional cytogenetic methods (direct preparation) is therefore possible, as opposed to second-trimester amniocentesis. The results of the test are available earlier to the patient, usually still in the first trimester. CVS might therefore be more acceptable than amniocentesis for women undergoing invasive testing for fetal karyotyping, especially if there is a high risk of chromosomal anomalies. Regarding the maternal risk of termination of pregnancy and psychological consequences, earlier diagnosis seems preferable. Several studies have also confirmed the women's preference of first-trimester CVS compared mid-trimester amniocentesis provided that both risks are equal [8]. Furthermore, chorionic villi are an excellent source for fetal DNA analysis which is needed for prenatal diagnosis of single-gene disorders, or for biochemical analysis which is used for diagnosis of certain metabolic diseases where the underlying genetic defect is unknown.

Basically, two different methods for CVS have found widespread use, the transcervical and the transabdominal sampling technique. In the transabdominal route, an 18- to 20-gauge needle is inserted transabdominally through the uterine wall directly into the placenta, avoiding the membranes and the amniotic cavity. Placental tissue is obtained through aspiration. Similar to amniocentesis, the 'free-hand' technique or a needle guide attached to the ultrasound transducer can be used. For the transcervical approach, different kinds of catheters or a bendable metal biopsy forceps can be used. The instrument is inserted through the cervix into the placenta under continuous ultrasound guidance, and a small amount of placental tissue is aspirated or extracted.

Several trials have compared the fetal loss rate between CVS and amniocentesis. Except for the European multicenter study [9], where the loss rate increased after CVS, these studies found that the fetal loss rate of first-trimester CVS is equal to second-trimester amniocentesis [10, 11], and remains in the magnitude of around 1%. The most probable explanation for the findings in the European study is the participation of many centers with little experience in the CVS technique. The number of repeat procedures was significantly higher in the CVS as compared to the amniocentesis group. This further suggests that the experience of the operator is more important for CVS as compared to conventional amniocentesis. The risk regarding the route of CVS seems to be similar with both techniques [12], although some studies suggest a slightly higher risk for the transcervical route [13].

In 1991, an association between early CVS (before 11 weeks of gestation) and severe transverse limb abnormalities was reported and subsequently confirmed [14]. The type of malformation is compatible with the chronologic pattern of limb development, suggesting that mechanisms such as procedure-associated hypoperfusion or embolization might be involved in the pathogenesis. Because limb development is essentially completed at 10 weeks of gestation, it is imperative that CVS is performed not before this time period, although this association was disputed later on in other registry-based reports [15, 16].

When comparing CVS to conventional amniocentesis, two further issues must be taken into account. Firstly, amniotic fluid can be used for analysis of the concentration of α-fetoprotein which is elevated in fetuses with spina bifida, encephalocele or gastroschisis. However, today these malformations can be diagnosed solely using high-resolution ultrasound in most cases. Secondly, CVS might reveal cells with different karyotype, a so-called mosaicism, which occurs in about 1% according to large series of CVS. This might be confined to the extrafetal tissue (placental-confined mosaicism which is associated with uniparental disomy and intrauterine growth restriction), or also concern the fetus [17]. In these cases it is therefore necessary to confirm or exclude mosaicism in the fetus by detailed ultrasound examination, fetal blood sampling or amniocentesis.

Placental Biopsy (Late CVS)

Rapid karyotyping by conventional cytogenetic analyses from chorionic villi does not need to be confined to the first trimester [18]. Although the mitotic index of cytotrophoblast cells decreases with increasing gestational age, karyotyping by direct preparation in addition to culture techniques is feasible. The main indications for the use of this method are ultrasound findings suspicious of an underlying chromosomal anomaly, and it can even be used in the third trimester up to term gestation [19]. It can be used as an alternative to fetal blood sampling and is, depending on the position of the placenta, usually technically easier.

Placental biopsy is performed by the transabdominal route with the same technique as described for early CVS, with a 20-gauge needle with or without a needle guide. The risk of fetal loss is estimated to be around 1%, although large series are lacking.

Early Amniocentesis

In the late 1980s, early amniocentesis was introduced with the aim of providing a tool for early karyotyping before 14 weeks of gestation. Amniotic fluid is obtained with a 20- to 22-gauge needle, as described for conventional amniocentesis. As the amniotic sac at this early gestational age is still small, concern has been raised about the possible consequences of withdrawal of sufficient amniotic fluid necessary for a successful cell culture. In order to obtain a maximum number of cells, amniofiltration has been proposed by several groups [20].

The first large comparative trial between CVS and early amniocentesis revealed a somewhat higher incidence of fetal loss and talipes equinovarus in the early amniocentesis group [21]. A subsequent Danish study was stopped early because of the finding of an increased incidence of talipes in the early amniocentesis group, whereas the fetal loss rate was not significantly different [22]. In the largest randomized study, the Canadian CEMAT trial, early amniocentesis was compared to late amniocentesis [23]. This study provided clear evidence that the fetal loss rate, postprocedural amniotic fluid leakage as well as the incidence of talipes is higher after early (11–13 weeks) as compared to second trimester (15–17 weeks) amniocentesis.

Fetal Blood Sampling (Cordocentesis)

Initially, fetal blood was obtained by a fetoscopic approach [24]. The umbilical cord insertion was directly visualized and punctured with a small needle for aspiration of fetal blood. This procedure was soon replaced by percutaneous ultrasound-guided fetal blood sampling, a technique similar to amniocentesis [25]. As with amniocentesis, the 'free-hand' technique or a needle guide can be used. Under direct ultrasound visualization, a 20- to 22-gauge needle is inserted into the umbilical vein, usually at the placental cord insertion site. The procedure can be performed either transplacentally or transamniotically, depending on the placental position and the location of the cord insertion.

The risk of fetal blood sampling depends on the indication for the procedure. In appropriately grown, structurally normal fetuses, the risk of fetal loss is 1–2%. However, this risk increases sharply if fetal blood sampling is performed in severely growth-restricted or abnormal fetuses.

Other Techniques of Invasive Testing

Other techniques for invasive fetal assessment are being used, such as cardiac puncture, puncture of intrahepatic veins, skin biopsy, liver biopsy, embryoscopy, urinary bladder puncture, pleurocentesis, etc. However, these procedures are only used for special indications because many of them have a significantly greater risk than the procedures described above, and extensive operator experience is mandatory.

Summary and Conclusions

To date, invasive testing remains the only possibility to definitively diagnose or exclude chromosomal disorders and genetic anomalies. Although several different techniques have been developed and assessed in large-scale trials over the last three decades, a procedure-related risk of fetal loss of around 1% remains the bottom-line of what must be regarded as the price which has to be paid for obtaining the diagnosis. Today's challenges regarding prenatal diagnosis are manyfold, but two issues seem to be primarily important: First, the development of valid measures for noninvasive risk assessment with the aim of reducing the number of unnecessary invasive procedures, and ultimately, the development of techniques for noninvasive diagnosis of chromosomal and genetic anomalies with high accuracy.

References

1 Valenti C, Schutta EJ, Kehaty T: Prenatal diagnosis of Down's syndrome. Lancet 1968;ii:220.

2 Hsu LYF: Prenatal diagnosis of chromosomal abnormalities through amniocentesis; in Milunsky A (ed): Genetic Disorders and the Fetus. Diagnosis, Prevention and Treatment, ed 3. Baltimore, Johns Hopkins University Press, 1992, pp 155–210.

3 Steele MW, Breg WR: Chromosome analysis of human amniotic fluid cells. Lancet 1966;i:383–385.

4 Bombard AT, Powers JF, Carter S, Schwartz A, Nitowsky HM: Procedure-related fetal losses in transplacental versus non-transplacental genetic amniocentesis. Am J Obstet Gynecol 1995;172: 868–872.

5 Tabor A, Philip J, Madsen M, Bang J, Obel EB, Norgaard-Pedersen B: Randomized-controlled trial of genetic amniocentesis in 4606 low-risk women. Lancet 1986;i:1287–1293.

6 Baird PA, Yee IM, Sadovnick AM: Population-based study of long-term outcomes after amniocentesis. Lancet 1994;344:1134–1136.

7 Midtrimester amniocentesis for prenatal diagnosis. Safety and accuracy. JAMA 1976;236:1471–1476.

8 Lippman A, Perry TB, Mandel S, Cartier L: Chorionic villi sampling: Women's attitudes. Am J Med Genet 1985;22:395–401.

9 Medical Research Council European trial of chorionic villus sampling. MRC working party on the evaluation of chorion villus sampling. Lancet 1991;337:1491–1499.

10 Multicentre randomised clinical trial of chorion villus sampling and amniocentesis. First report. Canadian Collaborative CVS-Amniocentesis Clinical Trial Group. Lancet 1989;i:1–6.

11 Smidt-Jensen S, Permin M, Philip J, Lundsteen C, Zachary JM, Fowler SE, Gruning K: Randomised comparison of amniocentesis and transabdominal and transcervical chrrionic villus sampling. Lancet 1992;340:1238–1244.
12 Brambati B, Terzian E, Tognoni G: Randomized clinical trial of transabdominal versus transcervical chorionic villus sampling methods. Prenat Diagn 1991;11:285–293.
13 Smidt-Jensen S, Philip J: Comparison of transabdominal and transcervical CVS and amniocentesis: Sampling success and risk. Prenat Diagn 1991;11:529–537.
14 Firth HV, Boyd PA, Chamberlain P, MacKenzie IZ, Lindenbaum RH, Huson SM: Severe limb abnormalities after chorion villous sampling at 56–66 days' gestation. Lancet 1991;337:762–763.
15 Froster UG, Jackson L: Limb defects and chrrionic villus sampling: Results from an international registry 1992–1994. Lancet 1996;347:489–494.
16 Schloo R, Miny P, Holzgreve W, et al: Distal limb deficiency following chorionic villus sampling? Am J Med Genet 1992;42:404–413.
17 Miny P, Basaran S, Pawlowitzki ICH, et al: Validity of cytogenetic analyses from throphoblast tissue throughout gestation. Am J Med Genet 1989;33:136–141.
18 Nicolaides KH, Soothill PW, Rodeck CH, Warren RC, Gosden CM: Why confine chorionic villus (placental) biopsy to the first trimester? Lancet 1986;i:543–544.
19 Holzgreve W, Miny P, Basaran S, Fuhrmann W, Beller FK: Safety of placental biopsy in the second and third trimesters. N Engl J Med 1987;317:1159.
20 Byrne DL, Marks K, Braude PR, Nicolaides KH: Amniofiltration in the first trimester: Feasibility, technical aspects and cytological outcome. Ultrasound Obstet Gynecol 1991;1:320.
21 Nicolaides KH, Brizot M, Patel F, Snijders R: Comparison of chorionic villus sampling and amniocentesis for fetal karyotyping at 10–13 weeks' gestation. Lancet 1994;344:435–439.
22 Sundberg K, Bang J, Smidt-Jensen S, et al: Randomised study of risk of fetal loss related to early amniocentesis versus chorionic villus sampling. Lancet 1997;350:697–703.
23 Randomised trial to assess safety and fetal outcome of early and mid-trimester amniocentesis. Canadian Early and Mid-trimester Amniocentesis Trial (CEMAT) Group. Lancet 1998;351:242–247.
24 Rodeck C, Campbell S: Umbilical-cord insertion as a source of pure fetal blood for prenatal diagnosis. Lancet 1979;i:1244–1245.
25 Daffos F, Capella-Pawlowsky M, Forestier F: Fetal blood sampling during pregnancy with use of a needle guided by ultrasound: A study of 606 consecutive cases. Am J Obstet Gynecol 1985;153:655–660.

Daniel V. Surbek, MD, Department of Obstetrics and Gynecology, University of Basel
Schanzenstrasse 46, CH–4031 Basel (Switzerland)
Tel. +41 61 325 91 74, Fax +41 61 325 92 94, E-Mail surbek@hotmail.com

Hahn S, Holzgreve W (eds): Fetal Cells and Fetal DNA in Maternal Blood. New Developments for a New Millennium. 11th Fetal Cell Workshop, Basel, 2000. Basel, Karger, 2001, pp 28–36

Fetal Cells in Maternal Blood: An Overview of the Basel Experience

Wolfgang Holzgreve, Xiao Yan Zhong, Carolyn Troeger, Isabelle Haari, Vivian Kiefer, Sinuhe Hahn

Department of Obstetrics and Gynecology, University of Basel, Switzerland

Introduction

In the early 1990s, our group, when still based in Münster, Germany, used the then novel method of magnetic cell separation (MACS) [1] for the successful enrichment of fetal erythroblasts from the blood of pregnant women [2, 3]. Our early experience with this technology indicated that erythroblasts could already be reliably enriched for early in pregnancy [4], and that by the use of appropriate fluorescent in situ hybridization (FISH) fetal aneuploidies could be detected prior to an invasive procedure [5]. This latter exciting development facilitated our participation as the only non-US member in the NIH-funded National Institutes of Child Health and Development Fetal Isolation Study (NIFTY) [6]. This facet is still been actively pursued in our laboratory.

In Basel we have made concerted efforts not only to optimize the enrichment procedure, but also to extend the scope of these investigations by including fetal single-gene disorders. For this purpose we have developed strategies for the effective analysis of single erythroblasts by single-cell PCR.

A chance finding during these initial exploratory investigations, which we first reported in 1994 [7], was that considerably more erythroblasts were present in the circulation of pregnant women with preeclampsia when compared to their healthy counterparts. This finding has provided the basis for several ongoing projects in our laboratory, including a large prospective study to investigate at what time point in pregnancy this anomaly occurs.

Enrichment of Fetal Erythroblasts by MACS

Akin to most researchers in the field, our laboratory has focused on the enrichment of fetal erythroblasts due to the various advantages these cells offer over potential other candidates, such as trophoblasts or fetal lymphocytes, in that they are abundant in early fetal blood, rare in the periphery of normal adults and are uni-nucleate [8, 9]. Furthermore by being short lived, unlike fetal lymphocytes, there is no risk of obtaining cells from a previous pregnancy. However, due to their scarcity, which is in the order of 1 in 10^6 to 1 in 10^7 maternal nucleated cells, specific enrichment strategies had to be developed to enable their retrieval from maternal blood samples.

Currently, the two most widely used technologies are fluorescent activated cell sorting (FACS) [10], a process which has been most explored in greatest detail by Bianchi et al. [11], and the magnetic equivalent MACS or magnetic cell sorting which was introduced for this task by our group [2]. Both technologies rely on the interaction of specific monoclonal antibodies with the target cell. For FACS, these antibodies are labelled with a fluorescent dye and the bound cells are identified by a computer-controlled laser. On the other hand, cell sorting by MACS employs antibodies which are conjugated with micromagnetic particles. The still most widely used cell surface antigen is CD71 (transferrin receptor) which is highly expressed on erythroblasts. The retention of suitably antibody-bound cells is facilitated by a strong magnetic field during which unbound cells are flushed out by several washing steps.

Even though our early studies, which included several large scale investigations of over 600 samples [12], had convinced us of the feasibility of using a MACS approach for the enrichment of fetal erythroblasts, there were several steps in the enrichment method which could be improved upon. One hindrance was the large maternal blood volume (40 ml) required for the combination of the triple-density gradient centrifugation step followed by two rounds of MACS separation. A further question was the specificity of the anti-CD71 antibodies for fetal erythroblasts. In a direct comparison using XY FISH, in which we examined 94 maternal blood samples enriched by our traditional triple-gradient/double MACS protocol in comparison with 332 samples which had been enriched by simple 1,077g/l Ficoll density gradient and a single round of MACS enrichment with anti-CD71, we determined that no significant loss in sensitivity or specificity resulted from the use of a more simple enrichment procedure [13]. This latter simpler enrichment procedure had the added advantage in that it permitted us to reduce the maternal blood requirement from 40 to 15 ml.

In a further series of experiments we determined that significantly more erythroblasts could be recovered by using higher Ficoll [14] or Percoll [15] density gradients, such as 1,119 g/l when compared to the standard of 1,077 g/l. These studies also indicated that an almost 5-fold greater yield of erythroblasts could be retrieved by using antibodies against more erythroblast-specific antigens, such as glycophorin A, a blood group antigen specific for the red cell lineage [14]. These experiments formed the basis of our later successful analysis of single fetal erythroblasts isolated by micromanipulation and examined by single cell PCR.

The NIH NIFTY Study

The encouraging findings in the early 1990s by our group and those of Elias and Bianchi [for review, see 8] that fetal cells could be used for the analysis of fetal aneuploidies were instrumental in convincing the NIH to fund a large-scale clinical study to test the efficacy of the new noninvasive methodology. For this purpose they initiated the so-called NIFTY study, the aim of which was to determine the relative efficacy of this emerging technology for the identification of fetal aneuploidies [6].

At the close of the first phase of this study early in 2000, close to 3,000 blood samples obtained from pregnant women who had an elevated risk of bearing a fetus with a chromosomal aberration were examined [16, 17]. Two of the participating laboratories (Basel and Jefferson) used MACS forms of enrichment, whereas two other groups used different FACS protocols (Boston and Houston). The preliminary analysis of these data, which have been presented at previous Fetal Cell Workshops or other international meetings, indicated that fetal aneuploidies could be detected with a higher specificity than current noninvasive methods [16, 17]. Evidence has also emerged that MACS enrichment procedures may be more efficient for the enrichment of fetal erythroblasts than comparable FACS ones. These issues will be resolved in the second phase of this study in which it is planned to use a common protocol for the 4 participating laboratories. Experiments performed during the NIFTY study have highlighted the possibility of examining multiple fetal chromosomes by multi-color FISH [18] or by repeatedly examining the same cell in multiple rounds of FISH [19, 20].

In our own studies, which include a published report on over 400 cases [13] and close to 1,000 cases entered into the central NIFTY database, we have focused mainly on the determination of fetal sex. Here we have obtained sensitivities close to 60% when using one XY-positive cell as being indicative that the women was pregnant with a son [13]. The specificity, however, was quite low, being in the order of 85%. This could be increased to over 95% if three or more XY-positive cells were observed, which was, however, accompanied by a concomitant drop in sensitivity to below 20%. This result implies that more emphasis will have to be placed on improving the efficiency of fetal cell recovery and also enhancing the detection of fetal erythroblasts. For this reason we are actively exploring the use of automated image analysis systems for the recognition of potential fetal cells, such as those described by Philip Cornett of Chemscan elsewhere in this volume, as these systems will permit the screening of several thousand cells.

A very important issue addressed by the NIFTY study is the question of whether pregnant women would feel coerced to undergo an involuntary prenatal diagnosis by the introduction of such new risk-free technologies [21]. In our own extension of this study in which we did not restrict our examination to high-risk pregnant women, but also questioned low-risk pregnant women, those who had undergone some form of assisted reproduction and those currently not pregnant, has indicated that no such reservation exists. On the contrary, many couples, especially those involved in an assisted reproductive technology program were found to welcome the introduction of such a noninvasive risk-free prenatal diagnostic methodology. This is understandable, as these couples are least willing to subject their long-desired baby to the risks associated with an invasive procedure.

In the extension of the NIH NIFTY study, several new developments will be addressed including the examination of fetal Mendelian genetic disorders by single-cell PCR, the use of free fetal DNA in maternal plasma, as well as the possibility of culturing fetal hemopoietic progenitor cells isolated from the maternal circulation. Our group is actively participating in both these new facets.

Characterization of Fetal Genetic Traits by Single-Cell PCR

The reports by Cheung et al. [22] as well as those by Sekizawa et al. [23, 24] which indicated that fetal genetic loci could be analyzed by examining fetal erythroblasts by PCR, inspired us to investigate this approach as well.

Although several groups, including Cheung et al. [22], have proposed the use of fetal hemoglobin molecules for the tentative identification of fetal erythroblasts, it has been our conviction that fetal erythroblasts cannot be reliably identified solely on the basis of a purported fetal antigen, but that the irrevocable identification can only be made by a genetic analysis. For this purpose we have investigated the use of microsatellite patterns, also termed short-tandem repeats (STRs). By the use of such highly polymorphic STR loci, we have shown that single fetal cells can be reliably distinguished from maternal cells [25]. These studies, however, also indicated that the PCR analysis of single erythroblasts is particularly prone to allele drop out, a phenomenon whereby only one allele of a pair is represented due to inadequate amplification of the other allele. To overcome this deficit, we developed a strategy whereby multiple fetal cells were analyzed singly. By these means we were able to show that the analysis of at least five single fetal erythroblasts would yield a diagnostic accuracy of close to 99% [25, 26].

Confident that we were able to retrieve single erythroblasts and be able to analyze them by single-cell PCR, we set out to address a rather contentious issue in the field, namely, what proportion of the erythroblasts in the maternal circulation was of fetal origin [27–29]. These reports, generated using FISH, had previously argued that the minority of these cells was of fetal origin. If this were indeed the case, then it might be more appropriate to seek a more abundant fetal cell type.

Due to the limitations and technical difficulty of the above approach using STRs, we decided to use an alternative system which could be used to rapidly examine a suitably large number of samples, namely the Y chromosome and the rhesus D gene in rhesus D pregnant women [30]. Both of these alleles are absent in the genome of the pregnant women examined, the PCR assay is relatively simple, and since only a single copy of each fetal gene is present, we do not need to worry about the problem of allele drop out. A further important aspect of this study was that it would permit us to explore the feasibility of using this approach in determining the fetal rhesus D genotype in pregnancies at risk for fetal hemolytic disease due to a rhesus D incompatibility. In our study, in which we examined single erythroblasts identified morphologically following May-Grünwald-Giemsa staining by a multiplex single cell PCR assay for both the fetal rhesus D status and sex, we were able to correctly determine both fetal loci in 14 of the 19 samples examined. In these 14 successfully examined cases half of the erythroblasts examined were determined to be of fetal origin [30]. In the remaining 5 cases we were unable to retrieve any erythroblasts.

We have recently extended these investigations in collaboration with Dr. E. Di Naro, who was a fellow in our laboratory, by examining the fetal β-globin genotype in pregnancies at risk for β-thalassemias [31]. In this study, in which the approach developed by Cheung et al. [22] was used, only potential fetal erythroblasts which had been identified by immunohistochemistry for ζ-globin were examined. To avoid possible contamination with maternal erythroblasts, we restricted our analysis to single micromanipulated erythroblasts and did not resort to cell pooling as used by Cheung et al. [22]. In our study [31], in which 6 maternal blood samples were examined, we were able to correctly identify the fetal genotype in all 6 cases. These experiments were all performed using manual micromanipulation with drawn glass microcapillaries, which was found to be a tedious and laborious task and is also associated with a relatively high degree of cell loss. For this reason the use of laser-mediated micromanipulation systems was investigated, the results of which are described by Zhong et al. elsewhere in this volume.

Disturbed Fetal Cell Traffic in Preeclampsia and Other Pregnancy-Related Disorders

A major focus in our laboratory has been to extend upon a novel finding that there are significantly more erythroblasts in the blood of pregnant women with preeclampsia when compared to normal healthy pregnant women [7, 12]. It is interesting to note that the first report of fetal cells being present in the maternal circulation was made at the turn of the 19th century by the German pathologist, Schmorl [32], who discovered trophoblasts in the lungs of women who had died of eclampsia [32]. Our first task was to establish whether these erythroblasts are of maternal or fetal origin. For this purpose a case-control study was performed in which only pregnancies with a singleton male fetus were examined [33]. The data of this study, which were also presented at a previous Fetal Cell Workshop held in London, showed that once again many more erythroblasts (approximately 5-fold) were present in the maternal circulation of preeclamptic pregnancies when compared to matched controls. However, as we were dealing with pregnancies with male fetuses, we were now able to determine what proportion of these cells were of fetal (male) origin by using FISH for the X and Y chromosomes. This experiment indicated that there were indeed approximately 5-fold more XY-positive fetal cells in the circulation of preeclamptic pregnant women than in the control cohort. This result showed that a large proportion of the erythroblasts were indeed derived from the fetus and that, hence, the traffic of fetal cells into the maternal circulation was significantly affected in preeclampsia [33, 34].

As the initiating lesions leading to preeclampsia are thought to occur early in pregnancy, probably at the time of implantation and subsequent modification of the maternal decidua [35], we next examined whether feto-maternal cell traffic is already affected early in those pregnancies which subsequently develop preeclampsia. For this purpose we performed a large-scale prospective study in which blood samples were collected at approximately 20 weeks of pregnancy, from which erythroblasts were enriched and enumerated [36]. These results were then compared to the subsequent pregnancy outcome. Of the 97 pregnant women examined, 72 had normal uncomplicated deliveries whereas 15 pregnancies were affected by preeclampsia. Interestingly, our study showed that the number of erythroblasts was already significantly elevated by a factor of almost 5-fold early before the onset of disease symptoms in those pregnancies which later developed preeclampsia. We did not observe any alteration in erythroblast numbers in 10 pregnancies affected by fetal growth retardation in the absence of preeclampsia [36].

Intrigued by our observation regarding altered fetal cell traffic in preeclampsia, we investigated whether a similar phenomenon occurs in other pregnancy-related disorders. In these various examinations, one extreme instance was noted in a case of polyhydramnios, in which an incredible 50-fold increase in erythroblast numbers was seen when compared to gestational age-matched healthy controls [37]. As this woman was pregnant with a male fetus, several of these erythroblasts could be isolated by micromanipulation and they were examined by single cell PCR for the presence of the Y chromosome. Astonishingly, we determined that almost 50% of these cells were of fetal origin. Although this is an interesting case report, it remains to be established whether such incredible elevations in fetal cell traffic are generally associated with polyhydramnios.

Conclusions

Although the original intention of our studies was to explore the use of a new enrichment method, MACS, for the retrieval of rare fetal erythroblasts and their subsequent analysis by FISH to detect fetal aneuploidies, this rather narrow aim was soon broadened by the development of new molecular biological techniques, such as single PCR [38], but also by chance, in that we observed preeclampsia to be associated with a disturbance in feto-maternal cell trafficking [34]. In this manner, it is to be expected that the traditional view of fetal cells in maternal blood, as a potentially useful tool for noninvasive prenatal diagnosis, is to be further challenged by the recent discoveries implying persistent fetal cell microchimerism in certain autoimmune disorders, such as scleroderma [39]. These new facets will contribute to more in-depth research in the field and thereby, hopefully, ensure a greater degree of understanding of the role, if any, of fetal cells in maternal blood during pregnancy.

Acknowledgements

This work was supported in part by the Swiss National Science Foundation, grant number 3200-047112.96, and the NIH (USA), contract number N01-HD-4-3202.

References

1 Miltenyi S, Müller W, Weichel W, Radbruch A: High gradient magnetic cell seperation with MACS. Cytometry 1990;11:231–238.
2 Holzgreve W, Garritsen HS, Gänshirt-Ahlert D: Fetal cells in the maternal circulation. J Reprod Med 1992;37:410–418.
3 Gänshirt-Ahlert D, Burschyk M, Garritsen HS, Helmer L, Miny P, Horst J, Schneider HP, Holzgreve W: Magnetic cell sorting and the transferrin receptor as potential means of prenatal diagnosis from maternal blood. Am J Obstet Gynecol 1992;166:1350–1355.
4 Gänshirt D, Garritsen H, Miny P, Holzgreve W: Fetal cells in maternal circulation throughout gestation. Lancet 1994;343:1038–1039.

5 Gänshirt-Ahlert D, Börejesson-Stoll R, Burschyk M, Dohr A, Gerritsen HSP, Helmer E, Miny P, Velasco M, Walde C, Patterson D, Teng N, Bhat NM, Bieber MM, Holzgreve W: Detection of fetal trisomies 21 and 18 from maternal blood using triple gradient and magnetic cell sorting. Am J Reprod Immun 1993;30:194–201.
6 de la Cruz F, Shifrin H, Elias S, Simpson JL, Jackson L, Klinger K, Bianchi DW, Kaplan SH, Evans MI, Holzgreve W, et al: Prenatal diagnosis by use of fetal cells isolated from maternal blood. Am J Obstet Gynecol 1995;173:1354–1355.
7 Gänshirt D, Börjesoson-Stoll R, Burschyk M, Garritsen HSP, Neusser M, Smeets FWM, Velasco M, Walde C, Holzgreve W: Successful prenatal diagnosis from maternal blood with magnetic-activated cell sorting. Ann NY Acad Sci 1994;731:103–114.
8 Hahn S, Sant R, Holzgreve W: Fetal cells in maternal blood: Current and future perspectives. Mol Hum Reprod 1998;4:515–521.
9 Holzgreve W, Hahn S: Fetal cells in cervical mucus and maternal blood. Baillières Best Pract Res Clin Obstet Gynaecol 2000;14:709–722.
10 Bianchi DW, Flint AF, Pizzimneti MF, Knoll JHM, Latt SA: Isolation of fetal DNA from nucleated erythrocytes in maternal blood. Proc Natl Acad Sci USA 1990;87:3279–3283.
11 Bianchi DW: Fetal cells in the maternal circulation: Feasibility for prenatal diagnosis. Br J Haematol 1999;105:574–583.
12 Gänshirt D, Smeets FW, Dohr A, Walde C, Steen I, Lapucci C, Falcinelli C, Sant R, Velasco M, Garritsen HS, Holzgreve W: Enrichment of fetal nucleated red blood cells from the maternal circulation for prenatal diagnosis: Experiences with triple density gradient and MACS based on more than 600 cases. Fetal Diagn Ther 1998;13:276–286.
13 Hahn S, Kiefer V, Brombacher V, Troeger C, Holzgreve W: Fetal cells in maternal blood: An update from Basel. Eur J Obstet Gynecol Reprod Biol 1999;85:101–104.
14 Troeger C, Holzgreve W, Hahn S: A comparison of different density gradients and antibodies for enrichment of fetal erythroblasts by MACS. Prenat Diagn 1999;19:521–526.
15 Smits G, Holzgreve W, Hahn S: An examination of different Percoll density gradients in combination with MACS for the enrichment of fetal erythroblasts. Arch Gynecol Obstet 2000;263:160–163.
16 de la Cruz F, Shifrin H, Elias S, Bianchi DW, Jackson L, Evans MI, Simpson JL, Holzgreve W, Klinger K: Low false-positive rate of aneuploidy detection using fetal cells isolated from maternal blood. Fetal Diagn Ther 1998;13:380.
17 Bianchi DW, Simpson JL, Jackson LG, Evans MI, Elias S, Holzgreve W, Sullivan LM, de la Cruz F: Fetal cells in maternal blood: NIFTY clinical trial interim analysis. Prenat Diagn 1999;19:994–995.
18 Bischoff FZ, Lewis DE, Nguyen DD, Murrell S, Schober W, Scott J, Simpson JL, Elias S: Prenatal diagnosis with use of fetal cells isolated from maternal blood: Five-color fluorescent in situ hybridization analysis on flow-sorted cells for chromosomes X, Y, 13, 18, and 21. Am J Obstet Gynecol 1998;179:203–209.
19 Zheng YL, Carter NP, Price CM, Colman SM, Milton PJ, Hackett GA, Greaves MF, Ferguson Smith MA: Prenatal diagnosis from maternal blood: Simultaneous immunophenotyping and FISH of fetal nucleated erythrocytes isolated by negative magnetic cell sorting. J Med Genet 1993;30: 1051–1056.
20 Zhen DK, Wang JY, Falco VM, Weber W, Delli-Bovi L, Bianchi DW: Poly-FISH: A technique of repeated hybridizations that improves cytogenetic analysis of fetal cells in maternal blood. Prenat Diagn 1998;18:1181–1185.
21 Zamerowski S, Lumley M, Arreola RA, Dukes K, Khan A, Sullivan L: The psychosocial impact on high-risk pregnant women of a noninvasive prenatal diagnostic test. Fetal Diagn Ther 1999;14: 125–126.
22 Cheung MC, Goldberg JD, Kan YW: Prenatal diagnosis of sickle cell anaemia and thalassaemia by analysis of fetal cells in maternal blood. Nat Genet 1996;14:264–268.
23 Sekizawa A, Kimura T, Sasaki M, Nakamura S, Kobayashi R, Sato T: Prenatal diagnosis of Duchenne muscular dystrophy using a single fetal nucleated erythrocyte in maternal blood. Neurology 1996;46:1350–1353.

24 Sekizawa A, Watanabe A, Kimura T, Saito H, Yanaihara T, Sato T: Prenatal diagnosis of the fetal RhD blood type using a single nucleated erythrocyte from maternal blood. Obstet Gynecol 1996;87: 501–505.
25 Garvin AM, Holzgreve W, Hahn S: Highly accurate analysis of heterozygous loci by single cell PCR. Nucleic Acids Res 1998;26:3468–3472.
26 Hahn S, Garvin A, Di Naro E, Holzgreve W: Allele drop-out can occur in alleles differing by a single nucleotide and is not alleviated by preamplification nor minor template increments. Genet Test 1998;2:351–355.
27 Slunga Tallberg A, el Rifai W, Keinanen M, Ylinen K, Kurki T, Klinger K, Ylikorkala O, Knuutila S: Maternal origin of nucleated erythrocytes in peripheral venous blood of pregnant women. Hum Genet 1995;96:53–57.
28 Slunga Tallberg A, el Rifai W, Keinanen M, Ylinen K, Kurki T, Klinger K, Ylikorkala O, Larramendy ML, Knuutila S: Maternal origin of transferrin receptor positive cells in venous blood of pregnant women. Clin Genet 1996;49:196–199.
29 Sohda S, Arinami T, Hamada H, Nakauchi H, Hamaguchi H, Kubo T: The proportion of fetal nucleated red cells in maternal blood: Estimation by FACS analysis. Prenat Diagn 1997;17:743–752.
30 Troeger C, Zhong XY, Burgemeister R, Tercanli S, Minderer S, Holzgreve W, Hahn S: Approximately half the erythroblasts in maternal blood are of fetal origin. Mol Hum Reprod 1999;5:1162–1165.
31 Di Naro E, Ghezzi F, Vitucci A, Tannoia N, Campanale D, D'Addario V, Holzgreve W, Hahn S: Prenatal diagnosis of beta-thalassaemia using fetal erythroblasts enriched from maternal blood by a novel gradient. Mol Hum Reprod 2000;6:571–574.
32 Schmorl G: Pathologisch-anatomische Untersuchungen über Präeklampsie. Leipzig, Vogel, 1893.
33 Holzgreve W, Ghezzi F, Di Naro E, Maymon E, Gänshirt D, Hahn S: Disturbed fetomaternal cell traffic in preeclampsia. Obstet Gynecol 1998;91:669–672.
34 Holzgreve W, Hahn S: Novel molecular biological approaches for the diagnosis of preeclampsia. Clin Chem 1999;45:451–452.
35 Roberts J, Redman C: Pre-eclampsia: More than pregnancy-induced hypertension. Lancet 1993; 341:1447–1451.
36 Holzgreve W, Li JC, Steinborn A, Külz T, Sohn C, Hodel M, Hahn S: Elevation in erythroblast count in maternal blood before the onset of preeclampsia. Am J Obstet Gynecol 2001;184:165–168.
37 Zhong XY, Holzgreve W, Li JC, Aydinli K, Hahn S: High levels of fetal erythroblasts and fetal extracellular DNA in the peripheral blood of a pregnant woman with idiopathic polyhydramnios: Case report. Prenat Diagn 2000;20:838–841.
38 Hahn S, Zhong XY, Troeger C, Burgemeister R, Gloning K, Holzgreve W: Current applications of single-cell PCR. Cell Mol Life Sci 2000;57:96–105.
39 Nelson JL: Microchimerism and autoimmune disease. N Engl J Med 1998;338:1224–1225.

Prof. Wolfgang Holzgreve, Department of Obstetrics and Gynecology, University of Basel
Schanzenstrasse 6, CH–4031 Basel (Switzerland)
Tel. +41 325 90 12, Fax +41 325 93 99, E-Mail wholzgreve@uhbs.ch

Hahn S, Holzgreve W (eds): Fetal Cells and Fetal DNA in Maternal Blood. New Developments for a New Millennium. 11th Fetal Cell Workshop, Basel, 2000. Basel, Karger, 2001, pp 37–40

Fetal Cells in Maternal Blood at Sheba Medical Center, Israel: Analysis of Three Cell Types during the Three Trimesters of Pregnancy

Esther Guetta, Michal Simchen, Keren Mammon, Ayala Aviram-Goldring, Gad Barkai

Danek Gertner Institute of Human Genetics, Sheba Medical Center and Department of Human Genetics and Molecular Medicine, Sackler Faculty of Medicine, Tel Aviv University, Tel Aviv, Israel

Introduction

The presence of rare fetal cells in maternal blood has been unequivocally proven by several methods, at different time points during pregnancy and several different cell types have been studied with a varied array of methods [1]. The clinical importance of the quantitative analysis of fetal cells in maternal blood stems from the need to determine the optimal time during pregnancy for noninvasive prenatal testing. Moreover, it is not known whether the transfer of different cell types occurs through a common mechanism, at similar frequencies, in a given pregnancy. In an attempt to shed light on these issues we chose to focus our initial efforts on the study of three cell types: mature red blood cells (RBC), nucleated red blood cells (NRBC), and hematopoietic progenitor cells (CD34+ cells) in maternal blood during pregnancy.

Experimental Approach

Blood samples were collected from a total of 48 women in 3 groups: 1st trimester (10–13 weeks); 2nd trimester (15–25 weeks), and 3rd trimester (28–41 weeks). Cell populations were separated by triple-density centrifugation [2] followed by magnetic activated cell

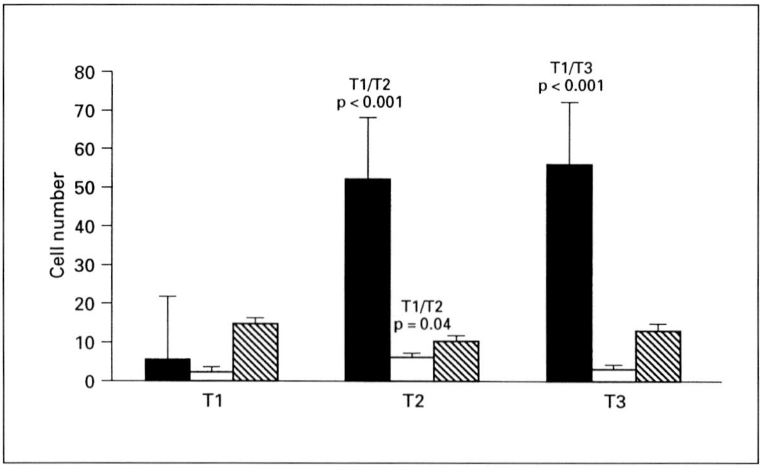

Fig. 1. Fetal cells in maternal blood in the three trimesters of pregnancy. Fetal NRBC (■), CD34+ cells (□) and fetal RBC (▧) in maternal blood in the 1st, 2nd and 3rd trimester samples (T1, T2, T3, respectively). Data shown represent the average number of fetal NRBC and CD34+ cells/10 ml maternal blood ± standard error and the average number of fetal RBC/1,000 total RBC ± standard error.

sorting with antibodies directed against CD71 and CD34 for enrichment of NRBC and CD34+ cells, respectively. Fetal RBC and NRBC were identified by immunocytochemistry (with an antibody directed against the γ-chain of fetal hemoglobin). CD34+ cells were identified by fluorescent in situ hybridization (FISH) in samples in which the fetus was male with probes for X and Y chromosomes.

Detection of Fetal Cells during Pregnancy

Analysis of fetal NRBC and CD34+ cell frequencies per 10 ml maternal blood revealed divergent patterns of distribution in the 3 sample groups (fig. 1). Both fetal NRBC and CD34+ cells increased from the 1st to 2nd trimester, by about 9- and 2-fold, respectively; however, the CD34+ cell frequency decreased in the 3rd trimester, whereas fetal NRBC remained at an increased level similar to that observed in the 2nd trimester. In contrast to the two nucleated cell types, we found a constant proportion of fetal erythrocytes in maternal blood relative to the total number of erythrocytes in the three trimesters (fig. 1).

In line with our own data, other groups have also found that CD71+ cell frequencies (measured by fluorescent activated cell sorting [FACS]) and fetal NRBC frequencies (measured by quantitative PCR and Kleihauer-Betke staining) increase with advancement of gestation [3, 4]. However, in another study in which fetal NRBC frequencies were studied with FACS and FISH, no change was found with relation to gestational age [5].

The use of fetal hemoglobin expression as a marker for fetal NRBC in maternal blood is not entirely specific and therefore it cannot be concluded that all the positive cells detected are indeed of fetal origin [6]. The success of fetal NRBC detection for noninvasive diagnosis will remain limited until a more specific method for fetal cell identification is developed. The application of antibodies raised against embryonic hemoglobin chains expressed in the 1st trimester has offered a partial solution to this difficulty [7].

Fetal CD34+ cells in maternal blood were analyzed in a limited number of studies [8, 9]. Fetal hematopoietic progenitor cells were detected in maternal blood and in some cases persisted after delivery, potentially complicating future noninvasive diagnoses. Moreover, a specific fetal CD34+ cell marker has yet to be identified, rendering fetal cell identification possible only if the fetus is male.

The detection of fetal hemoglobin-expressing mature (i.e. non-nucleated) erythrocytes in maternal blood was reported in early studies [10, 11]. The average frequency of these cells in maternal blood found in our study was similar to that found in the normal adult nonpregnant population [12]. We did not find an increase in fetal hemoglobin-positive cells with progression of pregnancy, as reported in earlier studies. This discrepancy is possibly a result of the small sample number in our study and/or the application of different methods.

It has not yet been determined whether the frequency of transfer of the different cell types is similar within a given pregnancy. We compared the frequency of fetal NRBC and CD34+ cell in individual samples and found similar levels for both cell types in some of our samples; however, a consistent, common trend in cell transfer was not detected.

Conclusions

Although the limitations currently involved in definitive fetal cell identification necessitate further research and development, the results of the present study indicate that simultaneous targeting of both NRBC and CD34+ cells in maternal blood samples from the 2nd trimester should generate a higher yield of fetal cells and thereby facilitate noninvasive prenatal diagnoses.

References

1 Martin WL, Durrant LG, Liu DT: Non-invasive fetal cell isolation from maternal blood. Br J Obstet Gynaecol 1998;105:576–583.

2 Bhat NM, Bieber MM, Teng NNH: One-step enrichment of nucleated red blood cells. J Immunol Methods 1993;158:277–280.

3 Bianchi DW, Zickwol GK, Yih MC, Flint AF, Geifman OH, Erikson MS, Williams JM: Erythroid-specific antibodies enhance detection of fetal nucleated erythrocytes in maternal blood. Prenat Diagn 1993;13:293–300.

4 Kuo PL: Frequencies of fetal nucleated red blood cells in maternal blood during different stages of gestation. Fetal Diagn Ther 1998;13:375–379.

5 Sohda S, Arinami T, Hamada H, Nakauchi H, Hamaguchi H, Kubo T: The proportion of fetal nucleated red blood cells in maternal blood: Estimation by FACS analysis. Prenat Diagn 1997;17: 743–752.

6 deGraaf IM, Jakobs ME, Leschot NJ, Ravkin I, Goldbard S, Hoovers JMN: Enrichment, identification and analysis of fetal cells from maternal blood: Evaluation of a prenatal diagnosis system. Prenat Diagn 1999;19:648–652.

7 Mesker WE, Ourkerk-V Velzen MCM, Oosterwijk, JC, Bernini LF, Golbus MS, Kanhai HHH, Van Ommen GJB, Tanke HJ: Two-colour immunocytochemical staining of gamma and epsilon type haemoglobin in fetal red cells. Prenat Diagn 1998;8:1131–1137.

8 Bianchi DW, Zickwolf GK, Weil GJ, Sylvester S, DeMaria MA: Male fetal progenitor cells persist in maternal blood for as long as 27 years postpartum. Proc Natl Acad Sci USA 1996;93:705–708.

9 Little MT, Langlois S, Wilson RD, Lansdorp P: Frequency of fetal cells in sorted subpopulations of nucleated red erythroid and $CD34^+$ hematopoietic progenitor cells from maternal peripheral blood. Blood 1997;89:2347–2358.

10 Popat N, Wood WG, Weatherall DJ, Turnbull AC: Pattern of F-cell production during pregnancy. Lancet 1977;ii:377–379.

11 Boyer SH, Belding TK, Margolet L, Noyes AN, Burke PJ, Bell WR: Variations in the frequency of fetal hemoglobin-bearing erythrocytes (F-cells) in well adults, pregnant women, and adult leukemics. Johns Hopkins Med J 1975;137:105–115.

12 Rochette J, Craig JE, Thein SL: Fetal hemoglobin levels in adults: Blood Rev 1994;8:213–224.

Dr. Esther Guetta, Danek Gertner Institute of Human Genetics, Sheba Medical Center
52621 Tel Hashomer (Israel)
Tel. +972 3 530 3974, Fax +972 3 530 2914, E-Mail guetta@post.tau.ac.il

Hahn S, Holzgreve W (eds): Fetal Cells and Fetal DNA in Maternal Blood. New Developments for a New Millennium. 11th Fetal Cell Workshop, Basel, 2000. Basel, Karger, 2001, pp 41–46

Nucleated Red Blood Cell Recovery from Maternal Blood with Non-Commercial Monoclonal Antibodies

F.V. Álvarez[a,b], *R. Alonso*[a], *A. Paz*[a], *J.H. Ladenson*[a], *B. Prieto*[a]

[a] Hospital San Agustín, Avilés, y
[b] Departamento de Bioquímica y Patología Molecular, Universidad de Oviedo, Asturias, España

Introduction

Prenatal diagnosis has advanced dramatically in the last three decades. In the 1950s genetic counseling was the only available option, but it was impossible to offer a definite diagnosis, not even through invasive methods. Currently, prenatal diagnosis of chromosome disorders relies on invasive tests, by amniocentesis or chorionic villus sampling, which have a low but not negligible risk of miscarriage. Therefore, these methods can only be offered to pregnant women considered at high risk of carrying a fetus with chromosome abnormalities.

The existence of fetal cells in maternal peripheral blood is already a well-established issue. The isolation and analysis of fetal cells present in maternal blood has become an attractive alternative method of noninvasive prenatal diagnosis and is being pursued by an increasing number of researchers. Fetal nucleated erythrocytes (NRBC) represent a promising target cell population since they have a full complement of nuclear genes and they are abundant in first-trimester fetal blood but rare in adult blood. Moreover, the erythrocyte line develops earlier in gestation than the white blood cell line, and NRBC express well-differentiated antigens, accessible to specific antibodies. They have a short life-span (25–35 days in adult blood) and therefore the isolation of NRBC from previous pregnancies is unlikely [1, 2].

However, fetal cells are so rare in the maternal circulation that a variety of technical approaches have had to be developed to enrich and isolate them. Single, double or triple density gradients are widely employed either as a first or as the only step in the enrichment of NRBC from maternal blood.

A variety of monoclonal antibodies (mAbs) have also been developed and used for fetal NRBC selection. Anti-CD71 mAb, which specifically recognizes the transferrin receptor, is the immunoreagent most commonly used in fetal NRBC isolation protocols [3]. The transferrin receptor is expressed on the NRBC surface from the earliest stages of maturation (BFU-E) until nuclear extrusion (reticulocytes) [4], but also on activated lymphocytes and monocytes [5, 6]. Despite not being specific to NRBC, the presence of fetal NRBC in maternal peripheral blood was successfully demonstrated for the first time by means of anti-CD71 mAb recognition [2]. Transferrin receptor reaches its highest expression in nucleated fetal cells during the first trimester of gestation, then decreasing with gestational age [7]. This expression is also increased in aneuploid fetuses [8], which allows the use of anti-CD71 mAb to isolate fetal NRBC from peripheral blood of pregnant women carrying chromosomally abnormal fetuses. However, since a considerable proportion of maternal mononuclear cells can also express the antigen CD71, leading to lower purity of fetal cells, the anti-CD71-based enrichment of fetal NRBC improves when maternal cells are eliminated by other means.

Our group has produced and characterized murine mAbs to human fetal liver NRBC. Some of these antibodies have been shown to recognize at least two different epitopes of the transferrin receptor on the surface of the NRBC [9] and they are being employed to isolate NRBC from maternal blood.

We have studied the distribution of the NRBC present in maternal blood in a double density gradient (1.077/1.107). The number of NRBC isolated from maternal blood samples was improved with a new protocol that involved the recovery of the cells followed by magnetic activated cell sorting (MACS) using our own mAbs. We have also observed that the delay in processing maternal blood samples influences the NRBC recovery from both interfaces of the double-density gradient.

Procedure

After informed consent, under a protocol approved by San Agustín Hospital, blood samples were collected from 62 pregnant women, between 10 and 16 weeks of gestation, into Vacutainer tubes containing EDTA anticoagulant (Becton Dickinson, UK).

A double density gradient was set up using Histopaque-1.107 (Histopaque-1.119 and Histopaque-1.077, 7:3) carefully overlayered with Histopaque-1.077 in a centrifuge tube (Iwaki, Japan), previously washed with PBS/EDTA/5% hSA, to prevent cell adhesion. Following this, blood diluted 1:1 in PBS/EDTA/0.1% hSA buffer was overlayered onto the

Histopaque-1.077 and the tubes were centrifuged at 440 *g* for 30 min at room temperature. Cells from the upper interface (plasma/1.077) as well as from the 1.077/1.107 interface were recovered and placed in separate tubes.

Two different mAbs (2E11.3 and 2B7.4), both directed against the transferrin receptor, were used for further positive MACS selection. Since our antibodies are not directly labeled with the microbeads, two consecutive incubations were carried out: first with the mAb and next with commercial magnetic microbeads (Miltenyi Biotec, GmbH, Germany).

The total number of positively selected cells was evaluated by a Neubauer counter chamber at a magnification of ×400, using phase-contrast optics. Cells were plated onto poly-*L*-lysine-coated slides using a cytocentrifuge (Shandon Cytospin 3, UK). Afterwards, slides were air-dried overnight before being fixed in methanol and stained with diaminobenzidine (DAB; Sigma, St. Louis, Mo., USA), to detect the presence of hemoglobin, as well as with hematoxylin, to counterstain the nuclei.

NRBC were identified according to their morphology and DAB staining.

Results and Discussion

The results obtained by using our mAbs with this approach are encouraging because NRBC were isolated from 61 of 62 consecutive maternal blood samples (98.4%). It is important to point out that the 2B7.4 mAb has succeeded in isolating NRBC from more than 200 consecutive samples.

The most common strategy for the NRBC isolation is density gradient centrifugation. The distribution of the NRBC population in density gradients depends mainly on the NRBC's stage of maturity, showing two peaks: one in the hypodense (1.065–1.085 g/ml) and the other in the normodense (1.075–1.093 g/ml) part of the gradient [10].

Different ways of target cell collection have been described when double-density gradients are used. Zheng et al. [11] as well as Wang et al. [12] collect the mononuclear cell layer from the interface of a Histopaque mixture of 1.109/1.077 (3:1) and the whole layer of Histopaque-1.077 together. Andrews et al. [13] observed that the most NRBC-enriched layer is the interface 1.109/1.077.

Our previous approach [9, 14] involved the recovery of cells from the 1.077/1.107 interface of a double Histopaque gradient. The main changes in the present protocol include prewashing centrifugation tubes with 5% hSA/PBS/EDTA, to prevent cell adhesion, and recovering cells from both interfaces, plasma/1.077 and 1.077/1.107, of the double density gradient. With these modifications, a significantly higher number of total NRBC was obtained (table 1), mainly from cells recovered from the upper interface of the gradient (median of NRBC: 23.4, range 0–1,560). The number of NRBC isolated from an average volume of 26 ml maternal blood was similar to that described by other authors in the literature, ranging from 0 to 2,000 in most of the cases, independent of the enrichment and isolation methods employed. Only Wachtel et al. [15] described an average number of

Table 1. NRBC isolated with the present protocol versus the previous one

	Previous protocol	Present protocol
Mean	44.41	194.63
Median	24	87.36
Minimum	3	4.3
Maximum	336	2,101.9

Mann-Whitney U test: $p < 0.01$.

NRBC which was tenfold higher than those of the other groups, by means of an original method of separation based on the electric charge of the cells (CFS).

Recovery of the less dense NRBC is accompanied by a higher contamination with maternal cells, mainly mononuclear leukocytes, which may complicate subsequent purification. In this case, however, contaminating cells did not constitute a problem for the identification of the NRBC purified by MACS with our mAbs 2E11.3 and 2B7.4.

The higher NRBC recovery obtained with the new protocol was encouraging enough to consider the upper layer as an additional potential source of NRBC, which had been previously discarded.

Since most of the maternal blood samples are not drawn at our laboratory, it was considered interesting to study the influence of the time elapsed between the blood drawing and its processing in the laboratory on the distribution of the NRBC in a double density gradient.

A clear tendency to lower NRBC recoveries was observed when processing 24-hour samples, although statistically significant differences were only shown at the plasma/1.077 interface (Wilcoxon signed rank test for paired samples, $p < 0.01$). Moreover, a higher number of contaminating cells was found when samples were processed after 24 h instead of within 3 h. This has also been observed by Oosterwijk et al. [16].

So far our results are encouraging because the yield and purity of the NRBC isolated from maternal blood could be further improved, for instance by an additional selective lysis step before the mAb incubation step, which would facilitate the NRBC-mAb interaction. Moreover, there is still a chance to enhance the number of NRBC from maternal blood by labeling our mAbs to the microbeads which would reduce the cell losses occurring with additional washes and centrifugations.

Minimization of contaminating cells would reduce the number of cytoslide preparations per sample, thus making the subsequent FISH analysis easier. It is also possible to optimize the enrichment conditions according to the elapsed time after maternal blood drawing by, for instance, modifying the centrifugation speed of the density gradient or by using different gradient densities.

Finally, our laboratory is now searching for the real percentage of fetal NRBC out of the total number isolated from maternal blood. We also consider it a priority to optimize the FISH technique in NRBC from maternal blood. It is well known that this technique has a high yield in models made with cord blood NRBC, but several groups have observed that the NRBC from maternal blood behave completely differently.

Acknowledgments

This project was supported by the Fondo de Investigación Sanitaria (98/1530), Fundación para el Fomento en Asturias de la Investigación Científica Aplicada y la Tecnología (PB-SAL98-02), and the Fundación Inocente. We thank all of the women who participated in this study, and Drs. Elena Vega, Ana Escudero, Paloma Díaz Velázquez, María Fernández Marmiesse, and all of the physicians who collaborated by sending us maternal blood samples. The antibodies used in this study are presently under option for exclusive license through the Washington University Center of Technology Management.

References

1 Lamvu G, Kuller JA: Prenatal diagnosis using fetal cells from maternal circulation. Obstet Gynecol Surv 1997;52:433–437.
2 Pertl B, Bianchi DW: First trimester prenatal diagnosis: Fetal cells in the maternal circulation. Semin Perinat 1999;23:393–402.
3 Price JO, Elias S, Wachtel SS, Klinger K, Dockter M, Tharapel A, Shulman LP, Phillips OP, Meyers CM, Shook D, Simpson JL: Prenatal diagnosis with fetal cells isolated from maternal blood by multiparameter flow cytometry. Am J Obstet Gynecol 1991;165:1731–1737.
4 Bianchi DW, Flint AF, Pizzimenti MF, Knoll JHM, Latt SS: Isolation of fetal DNA from nucleated erytrocytes in maternal blood. Proc Natl Acad Sci USA 1990;87:3279–3283.
5 Simpson JL, Lewis DE, Bischoff FZ, Elias S: Isolating fetal nucleated red blood cells from maternal blood: The Baylor experience 1995. Prenat Diagn 1995;15:907–912.
6 Larrick JW, Cresswell P: Modulation of cell surface iron transferrin receptors by cellular density and state of activation. J Supramol Struct 1979;11:579–586.
7 Zheng YL, Zhen DK, DeMaria MA, et al: Search for the optimal fetal cell antibody: Results of immunophenotyping studies using flow cytometry. Hum Genet 1997;100:35–42.
8 Thilaganathan R, Meher-Homji NJ, Nicolaides HK: Blood transferrin receptor expression in chromosomally abnormal fetuses. Prenat Diagn 1995;15:282–284.
9 Álvarez FV, Olander J, Crimmins D, Prieto B, Paz A, Alonso R, Porter S, Hess J, Crist RD, Landt Y, Ladenson J: Development, characterization and use of monoclonal antibodies made to antigens expressed on the surface of fetal nucleated red blood cells. Clin Chem 1999;45 (9):1614–1620.

10 Sitar G, Manenti L, Farina A, Lanati V, Mascheretti P, Forabosco A, Montanari L, Ascari E: Characterization of the biophysical properties of human erythroblasts as a preliminary step to the isolation of fetal erythroblasts from maternal blood for noninvasive prenatal genetic investigation. Haematologica 1997;82:5–10.
11 Zheng YL, Carter NP, Price CM, Colman SM, Milton PJ, Hackett GA, Greaves MF, Ferguson-Smith MA: Prenatal diagnosis from maternal blood: Simultaneous immunophenotyping and FISH of fetal nucleated erythrocytes isolated by negative magnetic cell sorting. J Med Genet 1993;30: 1051–1056.
12 Wang JY, Zhen DK Falco VM, Farina, Zheng YL, Delli-Bovi LC, Bianchi DW: Fetal nucleated erythrocyte recovery: Fluorescence activated cell sorting-based positive selection using anti-gamma globin versus magnetic activated cell sorting using anti-CD45 depletion and anti-gamma globin positive selection. Cytometry 2000;39:224–230.
13 Andrews K, Wienberg J, Ferguson-Smith MA, Rubinsztein DC: Enrichment of fetal nucleated cells from maternal blood: Model test system using cord blood. Prenat Diagn 1995;15:913–919.
14 Paz A, Prieto B, Alonso R, Cándenas M, Moína MJ, Porter S, et al: Aislamiento de eritroblastos fetales en sangre periférica materna mediante anticuerpos no comerciales. Prog Diagn Prenat 1999; 11:44–52.
15 Wachtel SS, Sammons D, Twitty G, Utermohlen J, Tolley E, Phillips O, Shulman LP: Charge flow separation: Quantification of nucleated red blood cells during pregnancy. Prenat Diagn 1998;18: 455–463.
16 Oosterwijk JC, Mesker WE, Ouwerkerk-van Velzen MCM, Knepflé CFHM, Wiesmeijer KC, van der Burg MJM, et al: Development of a preparation and staining method for fetal erythroblasts in maternal blood: Simultaneous immunocytochemical staining and FISH analysis. Cytometry 1998; 32:170–177.

Francisco V. Álvarez, Servicio de Análisis Clínicos, Hospital San Agustín
C/ Camino de Heros n° 4, E–33400 Avilés, Asturias (Spain)
Tel. +34 985 123014, Fax +34 985 123025, E-Mail falvarez@arrakis.es

Hahn S, Holzgreve W (eds): Fetal Cells and Fetal DNA in Maternal Blood. New Developments for a New Millennium. 11th Fetal Cell Workshop, Basel, 2000. Basel, Karger, 2001, pp 47–55

How to Enrich and Analyze Fetal Cells from Maternal Blood: 5-Year Experience with PCR, FISH and Cultivation

Harald Hohmann[a], *Susanne Michel*[a], *Wolfgang Reiber*[b], *Michael Günther*[c], *Uwe Claussen*[a], *Ferdinand von Eggeling*[a]

[a] Institute of Human Genetics and Anthropology, University of Jena;
[b] Private Gynecologist, Jena, and
[c] Department of Obstetrics, University of Jena, Germany

Introduction

Fetal nucleated erythrocytes in the maternal circulation are a valuable source for noninvasive prenatal diagnosis since their proportion in normal adult blood is low but abundant in the early fetus. There is no risk of obtaining cells from previous pregnancies because of their short half life, which is in contrast to lymphocytes with their potential to persist for more than 20 years [1]. Fetal nucleated erythrocytes are very rare in maternal blood (the ratio of fetal to maternal cells ranges between 1:10,000 and 1:1,000,000) [2]. Various enrichment methods to overcome this limitation exist. One of the main methods is density gradient centrifugation in different variations [3]. After this step the protocols rely on magnetic activated cell sorting (MACS) [4] or fluorescent activated cell sorting (FACS) [5] using monoclonal antibodies mainly against erythroid progenitors (e. g. CD71, glycophorin A, erythropoietin). Thereafter, the origin of the nucleated cells obtained from the blood of pregnant women has to be determined because it is important to deal with pure fetal material. The origin of enriched cells can be analyzed by two techniques. One possibility is PCR, especially a combination of single-cell random PCR [6] to amplify the whole genomic DNA at least 50-fold, followed by specific PCR-mediated polymorphism analysis to determine the origin of cells [7–9]. The DNA of a single cell can be of either fetal or maternal origin, but not both, as it would be likely that two or more cells are analyzed. Due to random PCR only a few fetal cells have to be found to assure reliable verification of genetic diseases.

On the other hand FISH is an established method [10, 11] demonstrating low percentages of chromosomal aneuploidies. Nevertheless, if the percentage of fetal cells is lower than the cutoff of the probe used, no reliable diagnosis is possible. Other methods described in the literature for detecting fetal cells are charged flow separation [12], ferrofluid suspension with a magnet [13], and flow sorting with intracytoplasmatic anti-fetal hemoglobin [14]. At the moment anti-fetal hemoglobin represents the best marker which differentiates between maternal and fetal nucleated erythrocytes even if the fetus is female and shows no genetic abnormalities. Herein we describe our experience with PCR and FISH for the detection of fetal cells. Additionally, the promising approach of cultivating fetal cells will be discussed.

Polymerase Chain Reaction

PCR is one method of choice to analyze enriched cells. It has been applied in the field of noninvasive prenatal diagnosis to amplify Y-specific sequences either from whole blood [15–17] or from single cells to elude maternal background [18]. A precise differentiation between maternal and fetal cells by PCR can only be obtained when single cells are used for analysis, because a single cell can be of either fetal or maternal origin. Two cells or more, however, could lead to an ambiguous result. The origin of the cell could be ascertained in the case of a male fetus by Y-specific primers or by highly variable microsatellites if the gender is unknown. The DNA amount of single cells would limit further experiments. In order to overcome this problem, we combined a type of random PCR called primer extension preamplification (PEP) [7] with a PCR using specific primers. With single-cell PEP, the DNA content can be amplified up to 50-fold. This offers the opportunity of multiple PCR testing, including the determination of the origin and sex of the samples, and the identification of several genetic defects causing Mendelian disorders.

In our study [19] we investigated 19 pregnant women and in total 134 single putative nucleated erythrocytes. For enrichment, triple-gradient centrifugation was used underlying 6 ml of diluted blood with 2 ml each of histopaque (Sigma) 1077, 1110 and 1119. Centrifugation was performed at 650 *g* for 30 min. Two different layers were isolated. The mononuclear layer, in which most of the leukocytes are located, was used to extract maternal DNA by standard methods [20]. The second layer containing nucleated erythrocytes was incubated for 15 min at 4°C with the monoclonal antibody CD71, which was coupled to magnetic beads (Miltenyi Biotech, Germany). Separation was performed with the MINI-MACS

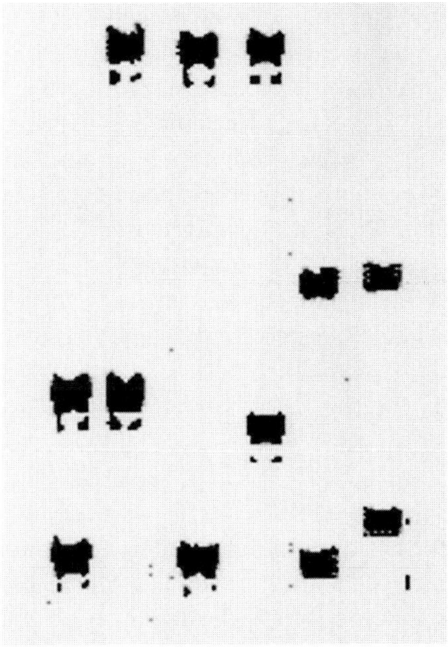

Fig. 1. PCR with primers flanking the tetranucleotide repeat D21S11. PCR products were electrophoresed and detected on an automatic infrared sequencer (LI-COR 4000, USA, NE). Lanes 1, 3 and 5 show products amplified from maternal DNA. In lanes 2, 4 and 6 the PCR product of a single PEP amplified cell is shown. As the allelic patterns match in only one allele with the maternal DNA, it can be stated that the cell was of fetal origin.

system (Miltenyi Biotech) as described by the manufacturer. The MACS-positive cell pellet was transferred to a Petri dish. The putative fetal cells were morphologically recognized with an inverse microscope (Axiovert, Zeiss, Germany). The collection of single cells was performed with extended siliconized Pasteur pipettes attached to a motor-driven micromanipulator. Each of the cells was transferred into separate caps and snap frozen in liquid nitrogen. A general DNA amplification of the single cells was performed by PEP as described elsewhere [7]. In order to detect the origin of the nucleated erythrocytes in a subsequent specific PCR, the allelic pattern of the mother was amplified in parallel with primers flanking highly polymorphic nucleotide repeats (short tandem repeats, STRs; D1S53, ACTBP2 and D21S11 [21, 22]) (fig. 1). In case of D21S11, the status of the ploidy for chromosome 21 can be ascertained in parallel [8, 23]. The amelogenin gene [24], located on chromosomes X and Y, was amplified and resulted in two different PCR products of 106 and 112 bp in length, respectively. When applying this system, differentiation between the maternal and fetal cells can be achieved when the fetus is male. Polymorphic repeat analysis revealed that 25% of the collected cells were of maternal origin, 26% of fetal origin, and 48% were uninformative. In cases with male fetuses the amelogenin system revealed 30% of the cells to be fetal nucleated erythrocytes, the remaining 70% being of maternal origin. This approach was used by other groups with lower numbers of cases analyzed [25, 26]. The results indicate a combination of random PCR and PCR-mediated polymorphism analysis on the DNA of single morphologically selected nucleated erythrocytes as a useful technique for noninvasive prenatal diagnosis. Until now no other technique except charged flow separation enables the detection of fetal cells in the 30% range.

Fluorescence in situ Hybridization

Fluorescence in situ hybridization (FISH) is another possibility to examine the enriched cells especially for the presence of the most common fetal chromosomal aneuploidies (13, 18, 21, X and Y). Since the studies by Klinger [27], who provided the first systematic analyses concerning the specificity and sensitivity of FISH, a lot of improvements for better results in the detection efficiency of fetal cells have been made. One major aspect for a successful FISH analysis is pretreatment of the cells. Different protocols were optimized [28, 29] to apply this technique to fetal nucleated erythrocytes sorted from the maternal circulation. A further aspect is the specificity of the hybridization probes, which plays a crucial role in the accuracy of prenatal diagnosis [30]. Beneath improvements in the hybridization efficiency of FISH probes, Zhen et al. [31] developed a method called poly-FISH. This protocol makes it possible to perform FISH at least nine times on the same interphase nucleus. A further progress in this field was the association of FISH and PCR, termed primed in situ (PRINS) labelling, combining the advantages of FISH-like visual assessment of in situ chromosome labelling and the powerful specificity and sensitivity of PCR for fetal cell detection [32]. Also other parameters can influence the efficiency of FISH. Recently, Brombacher et al. [33] described lower detection rates in blood samples treated with heparin than samples chelated with EDTA. Besides the enrichment of fetal cells using MACS leads to a higher specificity and less cell loss during FISH in contrast to FACS [34]. In another study Troeger et al. [35] compared different density gradients resulting in the conclusion that most erythroblasts were found in the buffy coat when gradients with higher density where used. All these protocols improve the FISH technique and lead to more sensitivity. But other factors like the scarcity of fetal cells and the poor hybridization efficiency of FISH probes limit the routine use of FISH.

In our study [36] we used FISH with chromosome X- and Y-specific probes for rapid testing of antibodies regarding their appropriateness to stain fetal nucleated cells. Therefore 20 ml of peripheral blood were collected from 40 women carrying a male fetus. Mononucleated cells enriched by density gradient were stained with antibodies previously not used for this purpose (2G5, 3F9, EPO-R and Flk1) and isolated with MACS. For FISH analysis the cell suspension was incubated at 37 °C for 20 min with hypotonic solution (0.075 *M* KCl), fixative (methanol:acetic acid 3:1) was added and the cells were centrifuged at 600 *g* for 5 min. Then the pellet was treated with fixative at –20 °C and after 5 min transferred onto slides. Dehydration was performed in an ethanol series (70, 95 and 100%). Afterwards fluorescence-labelled probes were hybridized. The specimen was denatured with formamide and 10 μl of the prewarmed probes were applied to the target. In our analysis, hybridization was performed using directly fluorescence-labelled centromere probes for the X- and Y-chromosome (Spectrum CEP Y green and Spectrum CEP X orange directly labelled probes; Vysis, Germany). The slides were sealed with rubber cement and placed in a 37 °C incubator overnight. After this incubation, the slides were treated with a rapid wash procedure using 0.4 × SSC for 2 min at 70 °C and 4 × SSC/Tween for 1 min at room temperature. Following dehydration in an ethanol series, cells were counterstained with DAPI (4,6-diamidino-2-phenyl-indol). Labelled cells were analyzed with a fluorescence microscope using the appropriate filter sets (Zeiss, Germany).

Our results showed that the tested antibodies, especially Flk1 (PharMingen, Europe) and EPO-R (MCA1636, Serotec Ltd., UK), were auspicious in the detection of fetal nucleated cells. Flk1 was originally a monoclonal antibody against the murine fetal liver kinase-1 receptor, also known as vascular endothelial growth factor receptor 2 (VEGF-R2) [37, 38]. EPO-R (Serotec, UK) is a monoclonal antibody against the erythropoietin receptor. Erythropoietin stimulates committed stem cells to increase the number of erythroid precursors, and accordingly the number of red blood cells entering the circulation. In conclusion, these two antibodies seem to be more successful in detecting fetal cells than the commonly used CD71 antibody, but further examinations are necessary to assure the results.

Cultivation

There is another possibility to overcome the problem of the scarcity of fetal cells in maternal blood by amplifying fetal progenitors in cultures, whereas PCR and FISH are mainly procedures to analyze enriched cells. Culture of just one fetal clonogenic cell could yield more than enough cells for genetic diagnosis, either molecular mutation testing or detection of aneuploidy by cytogenetic techniques. The existence of fetal clonogenic cells was demonstrated in different reports [28, 39, 40]. With regard to this approach for the enrichment of fetal nucleated cells, we investigated the possibility of culturing fetal erythroid progenitors. In our experiments, blood was used exclusively from pregnant women with known male fetuses. After triple-gradient centrifugation, two different fractions containing various progenitor cells were isolated and separately cultured in small Petri dishes using semisolid methylcellulose medium supplemented with growth factors like erythropoietin. After about 7 days all cultures showed multiple differentiated erythroid and myeloid cell clones. All clones of each fraction grown in one Petri dish were pooled and transferred into a PCR cap. Additionally, single erythroid clones were picked up from the remaining Petri dishes with extended Pasteur pipettes and each transferred into a PCR cap. From pooled clones as well as from single-cell clones, DNA was isolated by proteinase K digestion. Some clones were successfully used for chromosome preparation. For differentiation between fetal and maternal cells, sex determination PCR with XY-specific primers (amelogenin) was performed.

As a result, Y-specific sequences indicating the presence of fetal cells were not detectable, either in single clones or in pooled clones. This confirms the maternal origin of all cultured cells. One reason might be that fetal nucleated erythrocytes with proliferation potential do not pass over into the maternal circulation. However, it is more likely that the chosen culturing system did not provide the specific conditions that are necessary for fetal progenitor cells to grow.

In further experiments we used the culture medium Chang C with known growth factors or supplemented it with further ingredients like vitamins B1, B2, B6, B12 and C. These factors are known to have an up to four times higher concentration in fetal blood than in maternal blood and their addition might so differentially support the growth of fetal progenitors. Also under these conditions the amelogenin-specific PCR detected no fetal cells. The specific supplement conditions did not result in a growth of fetal cells. Ingredients like Flt3, an additional member of the receptor-type tyrosine kinase class III that is expressed on early hematopoietic progenitor cells [41, 42] would possibly improve the fetal cell growth conditions.

Conclusion

As already mentioned the two main methods to analyze fetal cells are PCR and FISH analysis. Combining MACS enrichment and the collection of single cells followed by PEP and specific PCR made it possible to identify about 30% fetal cells. The disadvantage of single-cell PCR-based analysis is that the collection of single cells needs a lot of experience with regard to micromanipulation and transferring the cells to reaction tubes. Further on the random amplification of low DNA amounts equivalent to a single cell could lead to artifacts in PCR, especially if STRs were amplified (allelic drop out [6]). With these analysis techniques, PCR is certainly more appropriate for the genetic analysis of small mutations.

However, FISH is mainly restricted by its very low percentage of fetal cells identified. This becomes especially evident in our investigations where only 1–2% of the cells were found to be of fetal origin. Mainly numeric aberrations, but also translocations and larger deletions can be analyzed. Furthermore, in FISH the counting of fetal cells can be realized automatically. With this automatic analysis it could be possible to search even large cell numbers for fetal cells in an adequate time [43].

In general, both methods have their assets and drawbacks. But up to now it is very difficult to identify small numbers of fetal cells for analysis. This is the reason why the great challenge for the future will be to find more effective methods for the detection and isolation of fetal cells. Endeavors to find fetal cell-specific antibodies are in progress but they are still not good enough for routine use. Cultivation could be an alternative attempt to increase the rare number of fetal cells and raise their quantity for easier analysis. In our efforts to reach this target we want to create a selective medium to improve the growth conditions for fetal hematopoetic precursors. Cultivation in combination with single clone analysis or a further enrichment with specific antibodies followed by FISH or PCR, may be the upcoming tool for the noninvasive diagnosis of fetal cells from maternal blood.

References

1 Bianchi DW, Klinger KW, Vadnais TJ, DeMaria MA, Shuber AP, Skoletsky J, Midura P, Williams JM: Developement of a model system to compare cell separation methods for the isolation of fetal cells from maternal blood. Prenat Diagn 1996;16:289–298.

2 Hamada H, Arinami T, Kubo T, Hamaguchi H, Iwasaki H: Fetal nucleated cells in maternal peripheral blood: Frequency and relationship to gestational age. Hum Genet 1993;91:427–432.

3 Oosterwijk JC, Mesker WE, Tanke HJ: Detection of fetal erythroblasts in maternal blood by one-step gradient enrichment and immunocytochemical recognition. Early Hum Dev 1996;47:95–97.

4 Ganshirt Ahlert D, Burschyk M, Garritsen HS, Helmer L, Miny P, Horst J, Schneider HP, Holzgreve W: Magnetic cell sorting and the transferrin receptor as potential means of prenatal diagnosis from maternal blood. Am J Obstet Gynecol 1992;166:1350–1355.

5 Bianchi DW, Flint AF, Pizzimenti MF, Knoll JH, Latt SA: Isolation of fetal DNA from nucleated erythrocytes in maternal blood. Proc Natl Acad Sci USA 1990;87:3279–3283.

6 von Eggeling F, Spielvogel H: Applications of Random PCR. Cell Mol Biol 1995;41:653–679.

7 Zhang L, Cui X, Schmitt K, Hubert R, Navidi W, Arnheim N: Whole genome amplification from a single cell: Implications for genetic analysis. Proc Natl Acad Sci USA 1992;89:5847–5851.

8 Mansfield ES: Diagnosis of Down syndrome and other aneuploidies using quantitative polymerase chain reaction and small tandem repeat polymorphisms. Hum Mol Genet 1993;2:43–50.

9 von Eggeling F, Michel S, Günther M, Schimmel B, Claussen U: Determination of the origin of single nucleated cells in maternal circulation by means of random PCR and a set of length polymorphisms. Hum Genet 1997;99:266–270.

10 Gänshirt-Ahlert D, Borjesson Stoll R, Burschyk M, Dohr A, Garritsen HS, Helmer E, Miny P, Velasco M, Walde C, Patterson D, et al: Detection of fetal trisomies 21 and 18 from maternal blood using triple gradient and magnetic cell sorting. Am J Reprod Immunol 1993;30:194–201.

11 Simpson JL, Elias S: Isolating fetal cells from maternal blood. Advances in prenatal diagnosis through molecular technology. JAMA 1993;270:2357–2361.

12 Wachtel SS, Sammons D, Manley M, Wachtel G, Twitty G, Utermohlen J, Phillips OP, Shulman LP, Taron DJ, Muller UR, Koeppen P, Ruffalo TM, Addis K, Porreco R, Murata Collins J, Parker NB, McGavran L: Fetal cells in maternal blood: Recovery by charge flow separation. Hum Genet 1996;98:162–166.

13 Martin B, Bruce J, Smith N, Liu D, Durrant L: A magnetic colloid system for isolation of rare cells from blood for fish analysis. Prenat Diagn 1997;17:1059–1066.

14 Zheng YL, Demaria M, Zhen D, Vadnais TJ, Bianchi DW: Flow sorting of fetal erythroblasts using intracytoplasmic anti-fetal haemoglobin: Preliminary observations on maternal samples. Prenat Diagn 1995;15:897–905.

15 Lo YM, Patel P, Sampietro M, Gillmer MD, Fleming KA, Wainscoat JS: Detection of single-copy fetal DNA sequence from maternal blood (letter). Lancet 1990; 335:1463–1464.

16 Kao SM, Tang GC, Hsieh TT, Young KC, Wang HC, Pao CC: Analysis of peripheral blood of pregnant women for the presence of fetal Y chromosome-specific ZFY gene deoxyribonucleic acid sequences. Am J Obstet Gynecol 1992;166:1013–1019.

17 Wachtel S, Elias S, Price J, Wachtel G, Phillips O, Shulman L, Meyers C, Simpson JL, Dockter M: Fetal cells in the maternal circulation: Isolation by multiparameter flow cytometry and confirmation by polymerase chain reaction. Hum Reprod 1991;6:1466–1469.

18 Takabayashi H, Kuwabara S, Ukita T, Ikawa K, Yamafuji K, Igarashi T: Development of non-invasive fetal DNA diagnosis from maternal blood. Prenat Diagn 1995;15:74–77.

19 von Eggeling F, Michel S, Günther M, Schimmel B, Claussen U: Determination of the origin of single nucleated cells in maternal circulation by means of random PCR and a set of length polymorphisms. Hum Genet 1997;99:266–270.

20 Grimberg J, Nawoschik S, Belluscio L, McKee R, Turck A, Eisenberg A: A simple and efficient non-organic procedure for the isolation of genomic DNA from blood. Nucleic Acids Res 1989;17: 8390.

21 Nishimura H, Kuzuya H, Kosaki A, Okamoto M, Kono S, Inoue G, Maeda I, Imura H: Monoclonal antibodies possibly recognize conformational changes in the human erythrocyte glucose transporter. Biochem J 1992;281:103–106.

22 Sharma V, Litt M: Tetranucleotide repeat polymorphism at the D21S11 locus. Hum Mol Genet 1992;1:67.

23 Pertl B, Yau SC, Sherlock J, Davies AF, Mathew CG, Adinolfi M: Rapid molecular method for prenatal detection of Down's syndrome. Lancet 1994;343:1197–1198.

24 Nakahori Y, Takenaka O, Nakagome Y: A human X-Y homologous region encodes 'amelogenin'. Genomics 1991;9:264–269.

25 Cheung MC, Goldberg JD, Kan YW: Prenatal diagnosis of sickle cell anaemia and thalassaemia by analysis of fetal cells in maternal blood. Nature Genet 1996;14:264–268.

26 Hahn S, Garvin AM, Di-Naro E, Holzgreve W: Allele drop-out can occur in alleles differing by a single nucleotide and is not alleviated by preamplification or minor template increments. Genet Test 1998;2:351–355.

27 Klinger KW: FISH: Sensitivity and specificity on sorted and unsorted cells. Ann NY Acad Sci 1994; 731:48–56.

28 Little MT, Langlois S, Wilson RD, Lansdorp PM: Frequency of fetal cells in sorted subpopulations of nucleated erythroid and CD34+ hematopoietic progenitor cells from maternal peripheral blood. Blood 1997;89:2347–2358.

29 Rodriguez-De AM, Palomino P, Jurado A, Sanz R, Ibanez MA, Fernandez MJ, Ayuso C, Diaz RJ, Lahoz C, Ramos C: Prenatal diagnosis on fetal cells obtained from maternal peripheral blood: Report of 66 cases. Prenat Diagn 1999;19:934–940.

30 Wei HJ, Su TH, Chien CL, Tzeng CR: Fluorescence in situ hybridization (FISH) as a method to detect aneuploid cells. Fetal Diagn Ther 1997;12:309–313.
31 Zhen DK, Wang JY, Falco VM, Weber W, Delli BL, Bianchi DW: Poly-FISH: A technique of repeated hybridizations that improves cytogenetic analysis of fetal cells in maternal blood. Prenat Diagn 1998;18:1181–1185.
32 Orsetti B, Lefort G, Boulot P, Andreo B, Pellestor F: Fetal cells in maternal blood: The use of primed in situ (PRINS) labelling technique for fetal cell detection and sex assessment. Prenat Diagn 1998;18:1014–1022.
33 Brombacher V, Kiefer V, Troeger C, Vial Y, Minderer S, Holzgreve W, Hahn S: Choice of anticoagulant can influence the analysis using fluorescence in situ hybridization of fetal cells enriched from maternal blood. Prenat Diagn 2000;20:257–259.
34 Samura O, Sekizawa A, Zhen DK, Falco VM, Bianchi DW: Comparison of fetal cell recovery from maternal blood using a high density gradient for the initial separation step: 1.090 versus 1.119 g/ml. Prenat Diagn 2000;20:281–286.
35 Troeger C, Holzgreve W, Hahn S: A comparison of different density gradients and antibodies for enrichment of fetal erythroblasts by MACS. Prenat Diagn 1999;19:521–526.
36 Hohmann H, Michel S, Reiber W, Gunther M, Claussen U, von Eggeling F: Applicability of four new antibodies for the detection of fetal nucleated cells out of maternal blood by FISH analysis. Fetal Diagn Ther 2001;16:52–56.
37 Shalaby F, Ho J, Stanford WL, Fischer KD, Schuh AC, Schwartz L, Bernstein A, Rossant J: A requirement for Flk1 in primitive and definitive hematopoiesis and vasculogenesis. Cell 1997;89: 981–990.
38 Kataoka H, Takakura N, Nishikawa S, Tsuchida K, Kodama H, Kunisada T, Risau W, Kita T, Nishikawa SI: Expressions of PDGF receptor alpha, c-Kit and Flk1 genes clustering in mouse chromosome 5 define distinct subsets of nascent mesodermal cells. Dev Growth Differ 1997;39:729–740.
39 Valerio D, Aiello R, Altieri V, Malato AP, Fortunato A, Canazio A: Culture of fetal erytroid progenitor cells from maternal blood for non-invasive prenatal genetic diagnosis. Prenat Diagn 1996;16: 1073–1082.
40 Valerio D, Altieri V, Antonucci FR, Aiello R: Characterization of fetal haematopoietic progenitors circulating in maternal blood of seven aneuploid pregnancies. Prenat Diagn 1997;17:1159–1169.
41 Findlay I, Matthews P, Quirke P: Multiple genetic diagnoses from single cells using multiplex PCR: Reliability and allele dropout. Prenat Diagn 1998;18:1413–1421.
42 Zeigler FC, Bennett BD, Jordan CT, Spencer SD, Baumhueter S, Carroll KJ, Hooley J, Bauer K, Matthews W: Cellular and molecular characterization of the role of the flk-2/flt-3 receptor tyrosine kinase in hematopoietic stem cells. Blood 1994;84:2422–2430.
43 Tanke HJ, Oosterwijk JC, Mesker WE, and Vrolijk J: Detection of 'rare event' fetal erythroblasts in maternal blood using automated microscopy. Early Hum Dev 1996;47:89–S93.

F. v. Eggeling, Institute of Human Genetics and Anthropology, University of Jena
Kollegiengasse 10, D–07740 Jena (Germany)
Tel. +49 3641 935526, Fax +49 3641 935582, E-Mail Fegg@mti-n.uni-jena.de

Hahn S, Holzgreve W (eds): Fetal Cells and Fetal DNA in Maternal Blood. New Developments for a New Millennium. 11th Fetal Cell Workshop, Basel, 2000. Basel, Karger, 2001, pp 56–62

Solid Phase Method for Enrichment of Fetal Cells from Maternal Blood

Reetta Vierumäki[a], *Kare Liimatainen*[a], *Martina Serlachius*[a], *Harriet von Koskull*[b], *Jim Schröder*[a]

[a] Division of Genetics, Department of Biosciences, University of Helsinki, and
[b] Laboratory of Prenatal Genetics, Department of Obstetrics and Gynaecology, Helsinki University Central Hospital, Helsinki, Finland

Introduction

Routine noninvasive prenatal diagnosis, based on the recovery of fetal cells from maternal peripheral blood, would benefit from a simple, inexpensive and efficient enrichment technique. Two enrichment methods are currently in use: sorting of antibody-labelled cells by fluorescence activated cell sorter (FACS) [1, 2] and magnetic activated cell sorter (MACS) [3, 4]. Neither of these techniques is yet suitable for routine prenatal testing. We have previously tried to overcome some of the problems associated with the use of antibodies (i.e. cross-reactivity, binding to Fc receptors) by using the ligand, transferrin, for the same purpose [5]. In this chapter we describe a simple and inexpensive enrichment method based on a coated solid phase and the transferrin receptor on the surface of nucleated red blood cells (NRBCs).

Materials and Methods

Eight umbilical cord blood samples (10–20 ml) from full-term deliveries and 4 blood samples (20 ml) from pregnant women were collected in EDTA tubes. Venous blood was obtained from pregnant women during the 10th and 11th week of gestation, prior to chorion villus sampling. All blood samples were processed within 24 h.

Table 1. Method for enrichment of nucleated red blood cells from cord blood and maternal blood samples

Cord blood sample or maternal blood sample	
↓	
1.119 density gradient centrifugation	
↓	
Separation by solid phase	
↓	↓
Cytospin	Extraction of genomic DNA
↓	↓
May-Grünwald-Giemsa staining	Nested X/Y-PCR

The blood was diluted 1:2 in phosphate-buffered saline (PBS) and layered on top of a single density gradient, Ficoll-Paque-1119® (Sigma, St. Louis, Mo., USA) [4]. After centrifugation at 600 *g* for 30 min at room temperature the two cell layers were recovered. Cells were washed with PBS and red blood cells lysed.

We tested the solid phase method for enrichment of NRBCs from umbilical cord blood. The following ligand and antibody combinations were tested: (1) plates coated with transferrin (Sigma); (2) plates coated with secondary antibody, rabbit α-mouse IgG2b (Zymed Laboratories Inc., San Francisco, Calif., USA) and cells labelled with a high-affinity α-CD71 monoclonal antibody of subclass IgG2b [6]; (3) plates coated with streptavidin (Sigma) and cells labelled with biotinylated transferrin (Sigma); (4) plates coated with streptavidin and cells labelled with biotinylated α-CD71 monoclonal antibody (Pharmingen, San Diego, Calif., USA), and (5) plates coated with streptavidin and cells labelled with biotinylated transferrin and biotinylated α-CD71 monoclonal antibody.

For enrichment of fetal NRBCs from maternal blood samples the plates were coated with streptavidin and cells labelled with biotinylated transferrin. The protocol is summarized in table 1.

Polystyrene Petri dishes (Corning Inc, Corning, N.Y., USA), 5 mm nominal diameter, were incubated overnight at 4 °C with 2 ml of 50 m*M* sodium bicarbonate (Sigma) containing antibodies or ligands. The plates were washed once with PBS, then 2 ml of blocking solution, 5% BSA in PBS (Sigma), was added to the plates. Plates were incubated for 1 h at 37 °C and washed once with PBS.

The cell suspension was incubated with antibodies or ligands for 30 min at 4 °C. Cells (5×10^6 to 2×10^7) were added in 2 ml of 5% FCS/PBS (Biological Industries, Israel) to the coated plates and plates were incubated for 30 min at 4 °C and for 15 min at 37 °C. Unattached cells were washed away with PBS and attached cells were eluated after incubation for 15 min at 37 °C with 1 × trypsin-EDTA (GibcoBRL, Germany).

Table 2. Enrichment of NRBCs from cord blood by solid phase with different ligands and antibodies

Plates coated with	Cells labelled with	Mean enrichment fold ± SD	Number of experiments	Total number of cells/plate
α-Mouse IgG2b	α-CD71 mAb (IgG2b)	11±0	1	1.8×10^7
Transferrin	–	8±0	1	5.0×10^6
Streptavidin	biotinylated transferrin	126±66	7	2.0×10^7
Streptavidin	biotinylated α-CD71 mAb	93±56	5	7.5×10^6
Streptavidin	biotinylated transferrin + biotinylated α-CD71 mAb	62±44	2	1.0×10^7

mAb = Monoclonal antibody

$$\text{Enrichment fold} = \frac{\text{NRBCs/nucleated cells after solid phase}}{\text{NRBCs/nucleated cells before solid phase}}$$

In experiments with cord blood, cytocentrifuge preparations were made (700 rpm for 12 min) and NRBCs were identified based on morphology and May-Grünwald-Giemsa staining (Merck, Germany).

In experiments with peripheral blood samples of pregnant women, genomic DNA was extracted from cells recovered from the solid phase [7]. A nested Y-specific PCR system detecting a single-copy gene, SRY, a testis-determining factor, present on the short arm of the Y chromosome [8], was carried out as described by van Wijk et al. [7]. A control gene sequence, a repeat present on the X chromosome, was amplified in the second-round polymerase chain reaction [7]. PCR products were analyzed in a 1.7% agarose gel (Sigma) and stained with ethidium bromide.

Results

Enrichment of NRBCs from Cord Blood

For the comparison of different ligand and antibody combinations, 8 cord blood samples divided into two parts (16 experiments) were studied. A mean of 0.39% (range 0.1–1.2%) NRBCs were found in samples before solid-phase separation after density centrifugation. Enrichment fold has been calculated by dividing the ratio of NRBCs to other nucleated cells after solid phase by the ratio of NRBCs to other nucleated cells before the solid phase. The yield of erythroblasts was highest when the solid phase was coated with streptavidin and the cells labelled with biotinylated transferrin (table 2; fig. 1). With streptavidin-coated plates and biotinylated α-CD71 monoclonal antibody (mAb)-labelled cells, the

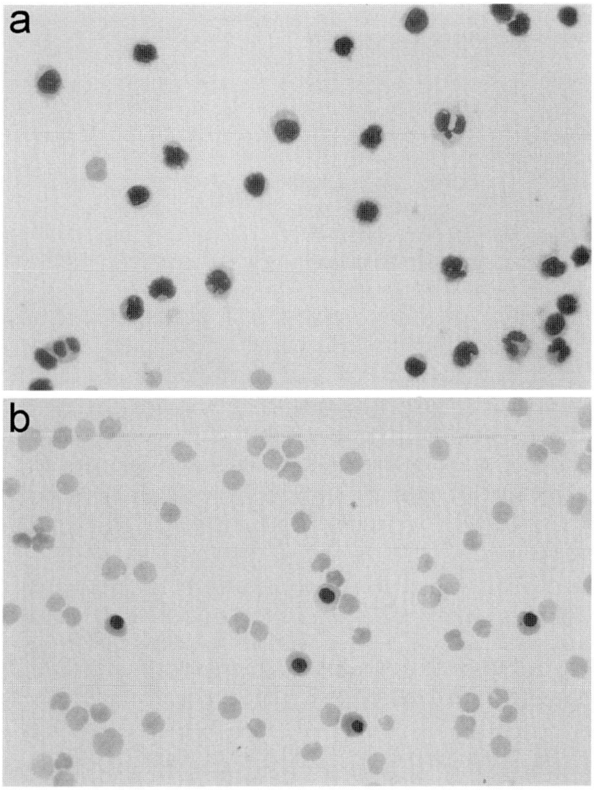

Fig. 1. ***a*** Cord blood cells before solid-phase enrichment. ***b*** Cord blood cells after solid phase enrichment. Nucleated red blood cells have been identified based on morphology and May-Grünwald-Giemsa staining. ×400.

enrichment of NRBCs was 93-fold. 'Cocktail' enrichment, in which cells were labelled both with biotinylated transferrin and biotinylated α-CD71 mAb, showed an enrichment fold of 62. The enrichments with rabbit α-mouse IgG2b-coated plates and α-CD71 mAb-labelled cells were less efficient than with streptavidin-coated plates. So were enrichments with plates coated only with transferrin.

Enrichment of NRBCs from Maternal Peripheral Blood

To enrich erythroblasts from maternal blood, the solid phase was coated with streptavidin and cells were labelled with biotinylated transferrin. Genomic DNA was extracted from solid phase-enriched cells and a specific and sensitive nested

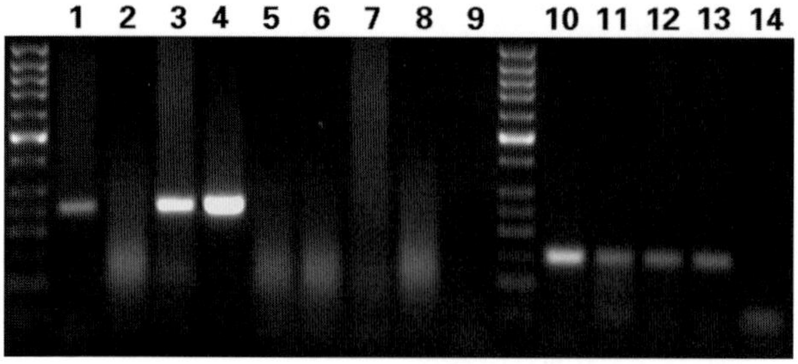

Fig. 2. PCR amplification of X- and Y-specific sequences from fetal cells enriched from maternal peripheral blood. Molecular weight marker is GeneRuler™ 50 bp DNA Ladder (Fermentas, Germany). Lanes 1–9 show results of Y-PCR analysis. Y-chromosome-specific fragment is 254 bp. Lane 1 is from patient 2. Lane 2 is a female control. Lane 3 is a 1:10,000 mixture of male and female DNA. Lane 4 is a male control. Lanes 5–8 are water controls for lanes 1–4. Lane 9 is a water control of second-round PCR. Lanes 10–14 show results of X-PCR analysis. The X-chromosome-specific fragment is 149 bp. Lane 10 is from patient 2. Lane 11 is a female control. Lane 12 is a 1:10,000 mixture of male and female DNA. Lane 13 is a male control. Lane 14 is a water control.

Table 3. Diagnosis of fetal sex by PCR on NRBCs isolated from a maternal blood sample by the solid-phase protocol compared with fetal karyotype

Patient No.	Gestation weeks + days	Karyotype of fetus (CVS)	Sex determined by PCR
1	10+2	46,XX	female
2	10+5	46,XY	male
3	10+1	46,XY	female
4	11+1	46,XX	female

X/Y PCR analysis was carried out. All samples were positive for the X chromosome (149 bp), indicating efficient isolation and transfer of DNA in all samples. The second PCR-round yields Y-specific product of 254 bp (fig 2). The karyotype of the fetus determined on chorionic villus cells and sex prediction by nested X/Y-PCR on recovered circulating fetal cells are shown in table 3. A male fetus was correctly identified in 1 of 2 women pregnant with a male fetus. No false-positive signals were detected in samples from 2 women pregnant with female fetuses.

Conclusion

We have developed a simple and inexpensive method to enrich NRBCs. Our data show that with this technique, a combination of streptavidin and biotinylated transferrin is most efficient for the enrichment of NRBCs from cord blood samples (table 2). This study confirms our previous results, which indicate that using a ligand instead of antibodies in the separation process is a viable and probably superior alternative [5].

Enrichment of fetal erythroblasts by MACS results in cell loss, disruption of cells and contamination by maternal cells. In the solid-phase method the target cells are recovered in a high yield with little cell loss and few contaminating maternal cells.

The efficacy of our method for enrichment of fetal cells from maternal blood is still under study. However, we have already extracted genomic DNA directly from cells, which have specifically bound to coated Petri dishes. We believe that this method decreases nonspecific cell loss inherent to other enrichment procedures currently in use. For amplification of Y chromosome-specific DNA we have chosen primers for the single-copy gene, SRY, present on the short arm of the Y chromosome [7]. A control gene sequence on the X chromosome has been amplified simultaneously with the nested Y-PCR to confirm the success of genomic DNA extraction. With this highly sensitive nested X/Y-PCR we are able to detect the presence of fetal cells in solid phase-enriched peripheral blood samples.

Acknowledgements

This study was financially supported by the Technology Development Center of Finland (TEKES). We would like to thank Dr. Erkki Koivunen from the Department of Biosciences, Division of Biochemistry, University of Helsinki, for advice concerning the solid phase method, and Dr. Pekka Katajisto from the Department of Biosciences, Division of Genetics, University of Helsinki, for the technical assistance with the PCR. We would also like to thank the staff at the Laboratory of Prenatal Genetics and the Delivery Unit, Department of Obstetrics and Gynaecology, Helsinki University Central Hospital, for helping with the collection of the blood samples.

References

1 Bianchi DW, Shuber AP, DeMaria M, Fougner AC, Klinger KW: Fetal cells in maternal blood: Determination of purity and yield by quantitative polymerase chain reaction. Am J Obstet Gynecol 1994;171:922–926.

2 Sekizawa A, Farina A, Zhen DK, Wang JY, Falco VM, Elmes S, Bianchi DW: Improvement of fetal cell recovery from maternal blood: Suitable density gradient for FACS separation. Fetal Diagn Ther 1999;14:229–233.

3 Zheng Y, Carter NP, Price CM, Colman SM, Milton PJ, Hackett GA, Greaves MF, Ferguson-Smith MA: Prenatal diagnosis from maternal blood: Simultaneous immunophenotyping and FISH of fetal nucleated erythrocytes isolated by negative magnetic cell sorting. J Med Genet 1993;30:1051–1056.
4 Troeger C, Holzgreve W, Hahn S: A comparison of different density gradients and antibodies for enrichment of fetal erythroblasts by MACS. Prenat Diagn 1999;19:521–526.
5 Serlachius M, von Koskull H, Wessman M, Schröder J: The use of transferrin for enrichment of fetal cells from maternal blood. Prenat Diagn 2000;20:407–410.
6 Nikinmaa B, Enns CA, Tonik SE, Sussman HH, Schröder J: Monoclonal antibodies to a purified human transferrin receptor. Scand J Immunol 1984;20:441–447.
7 van Wijk IJ, van Vugt JMG, Mulders M, Konst A, Weima SM, Oudejans CBM: Enrichment of fetal trophoblast cells from the maternal peripheral blood followed by detection of fetal deoxyribonucleic acid with a nested X/Y polymerase chain reaction. Am J Obstet Gynecol 1996;174:871–876.
8 Cui K, Warnes GM, Jeffrey R, Matthews CD: Sex determination of preimplantation embryos by human testis-determining-gene amplification. Lancet 1994;343:79–82.

Reetta Vierumäki, Division of Genetics, Department of Biosciences, University of Helsinki
Biocenter 2, Viikinkaari 5, PO Box 56, FIN–00014 Helsinki (Finland)
Tel. +358 9 19159688, Fax +358 9 19159079, E-Mail rvieruma@cc.helsinki.fi

Hahn S, Holzgreve W (eds): Fetal Cells and Fetal DNA in Maternal Blood. New Developments for a New Millennium. 11th Fetal Cell Workshop, Basel, 2000. Basel, Karger, 2001, pp 63–73

Fetal Nucleated Red Blood Cells in Maternal Blood during Pregnancy: First Czech Experience

Ilona Hromadnikova[a], *Nina Bendukidze*[d], *Sultan Karamanov*[e], *Lenka Sedlackova*[f], *Katerina Stechova*[a], *Bela Houbova*[a], *Miluse Mrstinova*[b], *Josef Kofer*[c], *David Stejskal*[g], *Milan Macek*[c], *Jan Vavrinec*[a]

[a] 2nd Department of Paediatrics,
[b] Department of Obstetrics and Gynaecology, and
[c] Department of Medical Genetics, 2nd Medical Faculty, University Hospital Motol, Prague;
[d] Department of Immunogenetics, Institute for Clinical and Experimental Medicine, Prague;
[e] Department of Medical Genetics, Masaryk Hospital, Usti nad Labem;
[f] Department of Immunology, KlinLab, Prague, and
[g] Department of Medical Genetics, GENNET, Prague, Czech Republic

Introduction

Prenatal genetic diagnosis is now available for woman at increased risk of carrying a fetus affected with a chromosomal aneuploidy or other genetically inherited diseases. However, all currently used invasive prenatal diagnosis methods, such as amniocentesis and chorionic villus sampling, carry procedure-related risks to the mother and fetus. Therefore, there is a great effort to develop a noninvasive techniques for prenatal diagnosis. The enrichment of fetal cells from maternal blood is an alternative promising source of fetal material for prenatal diagnosis [1–18]. Different types of fetal cells have been found in low frequencies in maternal peripheral blood [11, 19], such as trophoblasts [20–22], nucleated red blood cells (NRBCs) [23–25], lymphocytes [26, 27] and granulocytes. NRBCs are considered now as the main fetal cell candidate for diagnostic analysis [5, 10, 14, 15, 17–19, 25, 28–31]. They are a full complement of nuclear gene cells detectable early during pregnancy [32]. No problem with persistence into the next pregnancy was observed due to their short life and limiting replicating capacity [17, 18]. They express several unique antigens [33], such as the transferrin receptor (CD71) [3, 9, 25], erythropoietin receptor [34], thrombospondin receptor (CD36), glycophorin A (GYPA) [5], blood type and fetal liver surface antigens [35, 36] and HLA antigens [1, 16], that allow their enrichment from maternal circulation. However, even after enrichment the majority of NRBCs are of maternal origin [32, 37–39]. Another very important finding for future diagnostics revealed elevated numbers of fetal cells in aneuploid pregnancies [40]. Aneuploid fetal cells can be more easily enriched due to higher expression of surface markers such as CD71 [17, 41, 42]. Nucleated erythrocytes also produce unique fetal hemoglobin chains such as ζ and γ, that can be used as markers for their identification [43–45].

In this study we focused on the enrichment and detection of fetal erythroblasts circulating in maternal blood. NRBCs were enriched from maternal peripheral blood mononuclear cells by high-gradient magnetic cell separation (MACS) [3]. For effective isolation of fetal cells we combined depletion of maternal cells using anti-CD14 and anti-CD45 microbeads and positive selection using anti-CD71 microbeads [13]. We evaluated CD14–/CD45–/CD71+ cell quantity and the efficiency of enrichment in healthy women carrying 1 fetus and in aneuploid pregnancies. Sorted cells were analyzed for the presence of paternal alleles by simultaneous co-amplification and detection of low polymorphic HLA-DQA1 [13, 46], low-density lipoprotein receptor (LDLR), GYPA, hemoglobin G γ-globin (HBGG), D7S8 and group-specific component (GC) loci (AmpliType PM+DQA1 PCR Amplification and Typing Kit). Isolated cells were also analyzed by dual-color fluorescent in situ hybridization (FISH) with X- and Y-specific probes.

Material and Methods

Samples

NRBCs were enriched from 5–28 ml of peripheral blood (EDTA) from pregnant women between the 10th and 39th weeks of gestation. CD14–/CD45–/CD71+ cell numbers were analyzed in 136 healthy women carrying 1 fetus and in another 14 women with aneuploid pregnancies: trisomy 21 (n = 8), trisomy 18 (n = 2), 47,XXY (n = 3) and 47,XXX (n = 1).

Nucleated Red Blood Cells Enrichment

The mononuclear cells were isolated by Ficoll-Paque (Pharmacia, Uppsala, Sweden) density gradient (specific density of 1,077) centrifugation (400 *g* for 40 min at 20°C). To deplete maternal leukocytes peripheral blood mononuclear cells (1×10^7) were simultaneously incubated with anti-CD14 (20 µl) and anti-CD45 (20 µl) microbeads (mouse anti-human superparamagnetic beads conjugated monoclonal antibodies; Miltenyi Biotec, Germany) in PBS containing 0.5% bovine serum albumin (BSA), 2 m*M* EDTA and 0.01%

sodium azide (NaN_3) at 6 °C for 15 min. The labelled cells were washed in PBS and applied to the MACS-BS column for VarioMACS (Miltenyi Biotec). The CD14/CD45-negative fraction was stained with anti-CD71 microbeads (20 μl) in 80 μl PBS/BSA/NaN_3 at 6 °C for 15 min. The cells were then washed and sorted using MiniMACS. The number of live cells was evaluated at every enrichment step by a Neubauer counter chamber. Dead cells were excluded by trypan blue staining.

Flow Cytometry

All sorting steps were analyzed by three-color cytometric analysis with a Coulter EpicsXL flow cytometer. Cells were stained with 5 μl of anti-CD71-FITC-conjugated and anti-CD45-phycoerythrine (PE)-conjugated monoclonal antibodies (Immunotech, Prague, Czech Republic) and incubated for 30 min. Live cells were gated according to scatter with propidium iodide specifically staining for dead cells.

Extraction of DNA and Polymerase Chain Reaction

Genomic DNA was isolated by the standard protein salting-out method from maternal and paternal blood and by QIAmp Blood Kit (QIAGEN, Germany) from enriched NRBCs. The DNA concentration was measured at 260 nm on DU640B spectrophotometer (Beckman, USA). Paternal, maternal and fetal allele genotyping was carried out using the AmpliType PM+DQA1 polymerase chain reaction (PCR) amplification and typing kit (Perkin Elmer Applied Biosystems, USA) according to the manufacturer's instructions. The AmpliType PM+DQA1 kit simultaneously co-amplifies 5 AmliType PM loci including the LDLR (2 alleles), GYPA (2 alleles), HBGG (3 alleles), D7S8 (2 alleles), GC (3 alleles) and the HLA DQA1 locus (7 alleles). In some cases with very low concentrations of DNA, the primer-extension preamplification (PEP) technique was done before specific amplification using Operon RAPD 10-mer Kit A containing randomly selected nonspecific 10-base oligonucleotide primers [48].

FISH analysis

Cells were fixed with 3:1 (methanol:acetic acid) fixative for 30 min at room temperature. The cells were spun at 400 *g*, resuspended at 100 μl of fixative and dropped onto glass slides cleaned in ethanol for FISH analysis and stored at –20 °C. Dual-color FISH analysis was performed using Spectrum CEP Y green and Spectrum CEP X orange directly labelled probes (Vysis, USA) according to the supplier's instructions. The slides were washed in 70, 85 and 100% ethanol at room temperature for 1 min. Then 10 μl of probe solution was applied and the slides were placed to the GeneAmp In Situ PCR System 1000 (Perkin Elmer, Applied Biosystems, USA) with a program file for DNA denaturation (73 °C for 5 min) and following hybridization (42 °C overnight). The next day the slides were washed in post-hybridization solution (0.4 × SSC, pH 7.0–7.5 at 73 °C for 2 min) followed by another wash in 2 × SSC/0.1% NP-40, pH 7.0–7.5 at room temperature for 5–60 s. Finally 10 μl of DAPI II (4′,6-diamidino-2-phenylindole) counterstain (Vysis, USA) was applied and slides were stored in the dark prior signal enumeration. Cells were analyzed using an Olympus BX-50 microscope (Olympus, Japan) equipped with a narrow-band ultraviolet fluorescence cube for DAPI (U-MNU, 360–370 nm), a wide band blue fluorescence cube for FITC (U-MWB, 450–480 nm), a super wide-band green cube for rhodamine (U-MSWG, 480–550 nm) and triple band filter cube for DAPI (390–400 nm)/FITC (480–495 nm)/TRITC (550–570) (BX-TFC1).

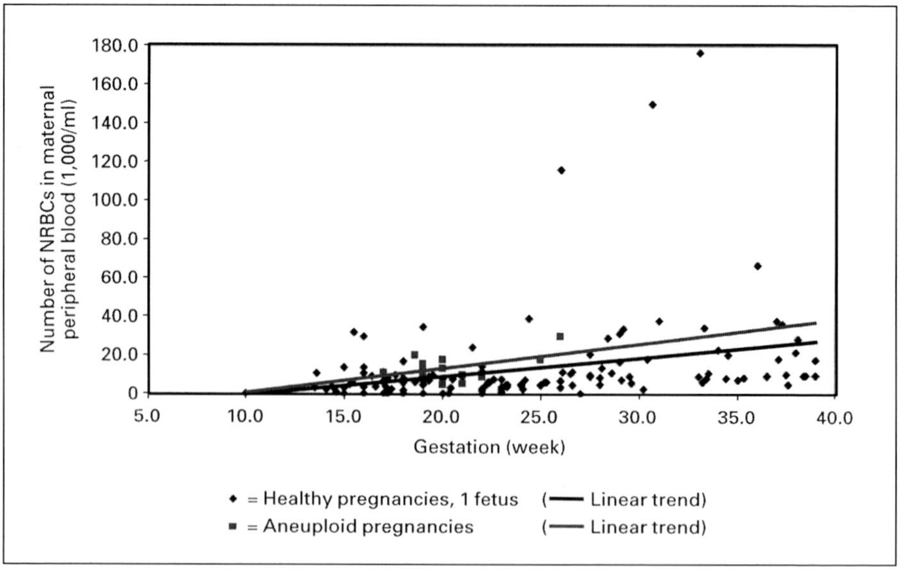

Fig. 1. Comparison of CD14–/CD45–/CD71+ cell frequencies in healthy women with aneuploid pregnancies carrying 1 fetus.

Results

Frequencies of CD14–/CD45–/CD71+ Cells Enriched from Maternal Blood

CD14–/CD45–/CD71+ cell numbers were analyzed in 136 healthy women carrying 1 fetus and in another 14 women with aneuploid pregnancies between 10 and 39 weeks of gestation. We observed that individual CD14–/CD45–/CD71+ cell numbers ranged from 1,000 to 176,500 (average 13,100)/ml of maternal blood. We found a steadily increase in cell frequencies with advancing gestation. We also observed higher CD14–/CD45–/CD71+ cell numbers in aneuploid pregnancies in comparison with a group of healthy women carrying 1 fetus (fig. 1).

Flow Cytometry Analysis

Peripheral blood mononuclear cells contained 72–96% (average 89%) CD45+ leukocytes and 0–1.8% (average 0.7%) CD45–/CD71+ cells. After the simultaneous CD45/CD14 cell depletion, the negative fraction contained 0.2–13.7% (average 3.5%) CD45+ cells. CD45–/CD71+ cells were enriched to 1.5–63% (average 18%) with enrichment rates ranged from 2- to 43-fold (average 21-fold). Using a double-MACS procedure, CD45–/CD71+ cells were enriched to 5.5–82.4% (average 56.2%) with a total average enrichment rate of 138-fold (4- to 526-fold). In the second MACS procedure the remaining CD45+ cells were again

Table 1. Efficiency of maternal cell depletion and nucleated red blood cell enrichment by double MACS procedure

	Peripheral blood mononuclear cells	1st MACS CD14/CD45 depletion	2nd MACS CD71+ selection
CD45–/CD71+ cells, %	0–1.8 (average 0.7)	1.5–63.0 (average 18.0)	5.5–82.4 (average 56.2)
Enrichment rate		2–43 (average 21)	4–526 (average 138)
CD 45+ cells, %	72.0–96.0 (average 89.0)	0.2–13.7 (average 3.5)	1.1–49.8 (average 15.4)

The number of live cells was evaluated at every enrichment step by a Neubauer counter chamber. Dead cells were excluded by trypan blue staining. All sorting steps were analyzed by three-color cytometric analysis with a CoulterEpicsXL flow cytometer. The enrichment rates were calculated as described by Miltenyi et al. [3].

enriched to 1.1–49.8% (average 15.4%; table 1). Flow cytometric histograms of NRBC separation are shown for 1 case in figure 2.

PCR Analysis of HLA-DQA1 and Polymarker System Paternally Inherited Loci

To identify paternally inherited alleles in a mixture of maternal and fetal cells, we used a PCR amplification and typing AmpliType PM+DQA1 kit. We successfully detected paternal alleles in 9 of 15 analyzed cases (60%). In 6 of 15 samples we did not observe any paternal allele amplification or the system was uninformative which means that the mother and father shared the same DNA markers. Comparing blue dot intensities within each locus, most of sorted CD14–/CD45–/CD71+ cells were of maternal origin. One informative case is presented in figure 3.

FISH Analysis

The results of dual-color FISH revealed that in 15 of 17 pregnant woman carrying a male fetus we were able to detect NRBCs of fetal origin with one X and one Y signal each. Only 1–13 fetal cells were detectable among 5–164 maternal cells. In average numbers, 3 fetal cells among 50 maternal ones were found (6%). In CD14–/CD45–/CD71+ cell samples enriched from 15 pregnant women carrying a female fetus, from 7 to 178 (average 56) fixed NRBCs with two X signals were detected. No difference in NRBC numbers between mothers carrying male and female fetuses was observed.

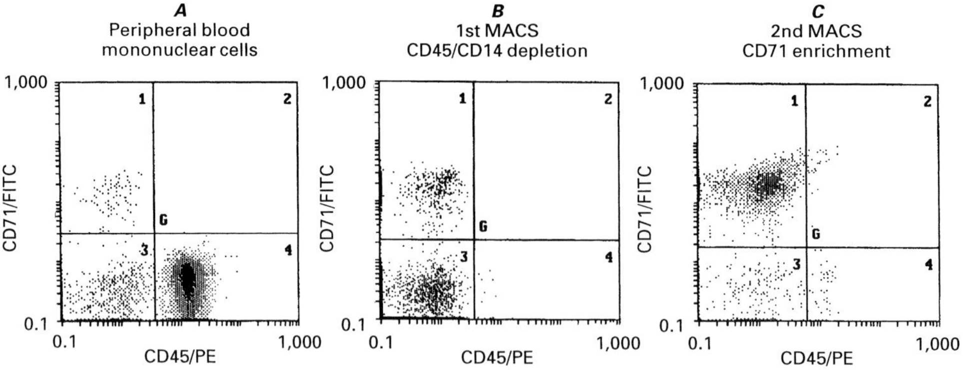

Region ID	ID	%	Count
G1	G	1.3	356
G2	G	0.0	12
G3	G	12.6	3,488
G4	G	86.1	23,794

Region ID	ID	%	Count
G1	G	14.4	955
G2	G	0	2
G3	G	85.1	5,660
G4	G	0.5	35

Region ID	ID	%	Count
G1	G	78.8	5,330
G2	G	1.8	123
G3	G	17.1	1,153
G4	G	2.3	155

Fig. 2. Flow cytometry histograms showing CD14–/CD45–/CD71+ cell enrichment from maternal circulation. Cells were stained with anti-CD71-FITC-conjugated and anti-CD45-PE-conjugated monoclonal antibodies. Live cells were gated according to the scatter with propidium iodide specifically staining for dead cells. Peripheral blood mononuclear cells contained 86.1% CD45+ leukocytes and 1.3% CD45–/CD71+ cells. After the simultaneous CD45/CD14 cell depletion the negative fraction contained 0.5% CD45+ cells. CD45–/CD71+ cells were enriched to 14.4%. Subsequent CD71 MACS positive selection enriched CD45–/CD71+ cells to 78.8% and the remaining CD45– cells to 2.3%.

Discussion

Fetal cells are very rare in maternal circulation with frequencies ranging from 10^{-4} to 10^{-8} [11, 19], which is too low to enable the detection of fetal alleles by sensitive molecular genetic methods such as PCR or FISH. Many methods to enrich fetal cells have been described. Currently FACS and MACS seem to be predominant [17]. The advantage of FACS is the achievement of a pure population with a defined phenotype due to simultaneous selection based on the CD71 surface expression, specific cell parameters as well as the presence of a fetal hemoglobin or any other specific marker [1, 19, 27, 30, 31]. Fetal cell enrichment may also be reached by an easy, fast and low-cost magnetic cell separation [3, 9, 10, 12,

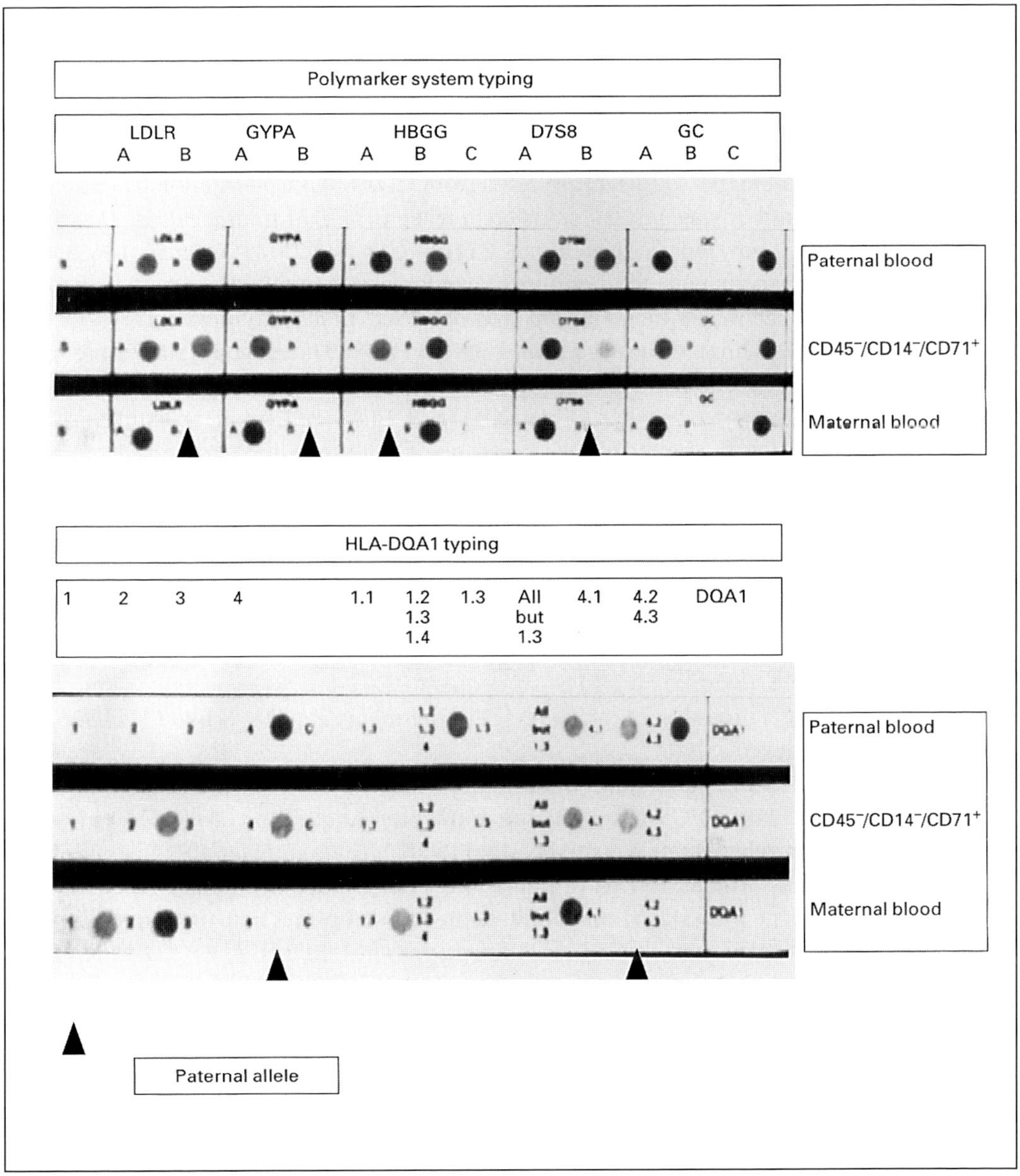

Fig. 3. PCR analysis of HLA-DQA1 and the polymarker system for paternally inherited loci. The interpretation of results by reading the pattern of blue dots on the AmpliType PM and HLA-DQA1 DNA probe strips determines the presence of maternal as well as fetal alleles in NRBCs enriched from the maternal circulation. In this case the results of PCR analysis revealed that the fetus inherited the paternal LDLR* B allele, GYPA* B allele, HBGG* A allele, D7S8* B allele and HLA-DQA1* 0501 allele.

13, 41]. In this study we summarized our experience with enrichment of fetal NRBCs from the maternal circulation using a combination of Ficoll-Paque density gradient centrifugation and high-gradient MACS [3]. For more effective isolation of fetal cells, maternal cells were first depleted using anti-CD14 and anti-CD45 superparamagnetic microbead-conjugated monoclonal antibodies [13]. Fetal cells were then positively selected from CD14–/CD45– fraction using a monoclonal antibody against the transferrin receptor (CD71 microbeads).

We analyzed frequencies of CD14–/CD45–/CD71+ cells in 136 healthy women carrying one fetus and in 14 women with aneuploid pregnancies between 10 and 39 weeks of gestation. We observed that individual CD14–/CD45–/CD71+ cell numbers ranged from 1,000 to 176,500 (average 13,100)/ml of maternal blood. We found a steady increase in cell numbers with advancing gestation. We also observed higher CD14–/CD45–/CD71+ cell frequencies in aneuploid pregnancies in comparison with a group of healthy women carrying 1 fetus. The combination of Ficoll density gradient centrifugation, maternal cell depletion and CD71-positive selection enabled us to enrich CD14–/CD45–/CD71+ cells in 146 of 150 pregnant women (97%) from the 13th week of gestation.

Nearly zero levels of CD71-positive cells (mean 0.7%) were found in maternal peripheral blood mononuclear cells. The CD45–/CD71+ cells were enriched to 1.5–63.0% (average 18%) after simultaneous maternal cell depletion and to 5.5–82.4% (average 56.2%) by subsequent CD71-positive selection. Using a double-MACS procedure a total average enrichment rate of 138-fold (4- to 526-fold) was achieved.

To identify paternally inherited alleles in a mixture of maternal and fetal cells, we used a PCR amplification and typing AmpliType PM+DQA1 kit which has been reliably and sensitively used to analyze mixtures of DNA samples containing as little as 150 pg of human DNA especially in forensic casework. The AmpliType PM+DQA1 kit simultaneously co-amplifies 6 distinct markers of low polymorphic loci including HLA DQA1 [13, 46], LDLR, GYPA, HBGG, D7S8 and GC. Here we show the successful detection of paternally inherited alleles in 9 of 15 analyzed cases (60%). In 6 cases it was impossible to demonstrate an allelic difference between the mother and fetus because the mother and father shared the same DNA markers or we could not observe any amplification of paternal alleles. Comparing blue dot intensities within each locus, most of sorted CD14–/CD45–/CD71+ cells were of maternal origin.

However, dual-color FISH with X- and Y-specific probes was carried out successfully in 88% of pregnant women carrying a male fetus and in 100% of women with a female fetus, the mean number of detected fetal cells isolated from maternal blood remaining very low. We were able to detect only 3 fetal cells among 50 maternal ones.

Our findings suggest that NRBCs in the maternal circulation are predominantly maternal in origin. For reliable prenatal diagnosis of fetal genetic abnormalities higher numbers of fetal cells would be desirable.

Acknowledgement

This work was supported by the Internal Grant Agency, Ministry of Health, Czech Republic, project No 4537-3 and by a grant from the 2nd Medical Faculty, Charles University in Prague, VZ No. 111300003.

References

1 Herzenberg LA: Fetal cells in the blood of pregnant women: Detection and enrichment by fluorescence-activated cell sorting. Proc Natl Acad Sci USA 1979;76:1453–1455.
2 Parks DR, Herzenberg LA: Fetal cells from maternal blood: Their selection and prospects for use in prenatal diagnosis. Methods Cell Biol 1982;26:277–295.
3 Miltenyi S, Muller W, Weichel W, Radbruch A: High gradient magnetic cell separation with MACS. Cytometry 1990;11:231–238.
4 Bianchi DW, Mahr A, Zickwolf GK, Houseal TW, Flint AF, Klinger KW: Detection of fetal cells with 47,XY,+21 karyotype in maternal peripheral blood. Hum Genet 1992;90:368–370.
5 Bianchi DW, Zickwolf GK, Yih MC, Flint AF, Geifman OH, Erikson MS, Williams JM: Erythroid-specific antibodies enhance detection of fetal nucleated erythrocytes in maternal blood. Prenat Diagn 1993;3:293–300.
6 Holzgreve W, Garritsen HSP, Gänshirt-Ahlert D: Fetal cells in maternal circulation. J Reprod Med 1992;37:410–418.
7 Holzgreve W, Garritsen HSP, Tercanli S, Miny P, Nippert I, Gänshirt D: Noninvasive prenatal diagnosis: Strategy for a clinical trial. Ann NY Acad Sci 1994;731:253–256.
8 Richkind KE: Advances and new directions in cytogenetic technology. Part II: Separation of fetal cells from maternal blood. Appl Cytogenet 1992;18:33–35.
9 Gänshirt-Ahlert D, Burschyk M, Garritsen HS, Helmer L, Miny P, Horst J, Schneider HP, Holzgreve W: Magnetic cell sorting and the transferrin receptor as potential means of prenatal diagnosis from maternal blood. Am J Obstet Gynecol 1992;166:1350–1355.
10 Gänshirt-Ahlert D, Borjesson-Stoll R, Burschyk M, Dohr A, Garritsen HS, Helmer E, Miny P, Velasco M, Walde C, Patterson D: Detection of Fetal trisomies 21 and 18 from maternal blood using triple gradient and magnetic cell sorting. Am J Reprod Immunol 1993;30:194–201.
11 Hamada H, Arinami T, Kubo T, Hamaguchi H, Iwasaki H: Fetal nucleated cells in maternal peripheral blood: Frequency and relationship to gestational age. Hum Genet 1993;91:427–432.
12 Zheng YL, Carter NP, Price CM, Colman SM, Milton PJ, Hackett GA, Greaves MF, Ferguson-Smith MA: Prenatal diagnosis from maternal blood: Simultaneous immunophenotyping and FISH of fetal nucleated erythrocytes isolated by negative magnetic cell sorting. J Med Genet 1993;30: 1051–1056.
13 Busch J, Huber P, Pluger E, Miltenyi S, Holtz J, Radbruch A: Enrichment of fetal cells from maternal blood by high gradient magnetic cell sorting (double MACS) for PCR-based genetic analysis. Prenat Diagn 1994;14:1129–1140.
14 Simpson JL, Elias S: Isolating fetal cells from maternal blood. Advances in prenatal diagnosis through molecular technology. JAMA 1993;270:2357–2361.
15 Simpson JL, Lewis DE, Bischoff FZ, Elias S: Isolating fetal nucleated red blood cells from maternal blood: The Baylor Experience 1995. Prenat Diagn 1995;15:907–912.
16 Steele CD, Wapner RJ, Smith JB, Haynes MK, Jackson LG: Prenatal diagnosis using fetal cells isolated from maternal peripheral blood: A review. Clin Obstet Gynecol 1996;39:801–813.

17 Hahn S, Sant R, Holzgreve W: Fetal cells in maternal blood: Current and future perspectives. Mol Hum Reprod 1998;4:515–521.
18 Bianchi DW: Current knowledge about fetal blood cells in the maternal circulation. J Perinat Med 1998;26:175–185.
19 Price PO, Sherman E, Wachtel SS, Klinger K, Dockter M, Tharapel A, Shulman L, Phillips OP, Meyers CM, Shook D, Simpson JL: Prenatal diagnosis with fetal cells isolated from maternal blood by multiparameter flow cytometry. Am J Obstet Gynecol 1991;165:1731–1737.
20 Schmorl G: Pathologisch-anatomische Untersuchungen über Präeklampsie. Leipzig, Vogel 1893.
21 Douglas GW, Thomas L, Carr M, Cullen NM, Morris R: Trophoblasts in the circulating blood during pregnancy. Am J Obstet Gynecol 1959;58:960–973.
22 Mueller UW, Hawes CS, Wright AE, Petropoulos A, DeBoni E, Firgaira FA, Morley AA, Turner DR, Jones WR: Isolating of fetal trophoblast cells from peripheral blood of pregnant women. Lancet 1990;336:197–200.
23 Kleinhauer E, Braun H, Betke K: Demonstration von fetalem Hämoglobin in den Erythrozyten eines Blutausstriches. Klin Wochenschr 1957;35:637–638.
24 Clayton EM, Feldhaus WD, Whiteacre FE: Fetal erythrocytes in the maternal circulation of pregnant women. Obstet Gynecol 1964;23:915–919.
25 Bianchi DW, Flint AF, Pizzimenti MF, Knoll JHM, Latt SA: Isolation of fetal DNA from nucleated erythrocytes in maternal blood. Proc Natl Acad Sci USA 1990;87:3279–3283.
26 Walknoska J, Conte FA, Grumbach, MM: Practical and theoretical implication of fetal/maternal lymphocyte transfer. Lancet 1969;i:1119–1122.
27 Iverson GM, Bianchi DW, Cann HM, Herzenberg LA: Detection and isolation of fetal cells from maternal blood using the fluorescence-activated cell sorter (FACS). Prenat Diagn 1981;1:61–73.
28 Chueh J, Globus MS: Prenatal diagnosis using fetal cells in the maternal circulation. Semin Perinatol 1990;14:471–482.
29 Bianchi DW, Stewart JE, Garber MF, Lucotte G, Flint AF: Possible effect of gestational age on the detection of fetal nucleated erythrocytes in maternal blood. Prenat Diagn 1991;11:523.
30 Price JO, Elias S, Wachtel SS, Klinger K, Dockter M, Tharapel A, Shulman LP, Philips OP, Meyers CM, Shook D, Simpson JL: Prenatal diagnosis using fetal cells isolated from maternal blood by multiparameter flow cytometry. Am J Obstet Gynecol 1991;165:1731.
31 Wachtel S, Elias S, Price J, Wachtel G, Philips O, Shulman L, Meyers C, Simpson JL, Dockter M: Fetal cells in the maternal circulation: Isolation by multiparameter flow cytometry and confirmation by polymerase chain reaction. Hum Reprod 1991;6:1466.
32 Gänshirt D, Garritsen HSP, Miny P, Holzgreve W: Fetal cells in maternal circulation throughout gestation. Lancet 1994;343:1038–1039.
33 Alter BP: Biology and erythropoiesis. Ann NY Acad Sci 1994;731:36–47.
34 Valerio D, Aiello R, Altieri V: Isolation of fetal erythroid cells from maternal blood based on expression of erythropoietin receptors. Mol Hum Reprod 1997;3:451–455.
35 Savion S, Carp H, Shepshelovich J, Irlin J, Kostykov M, Fein A, Toder V: Use of antibodies against the human antigen of erythroblasts for the detection of nucleated erythrocytes in the maternal circulation. Biol Neonate 1997;71:126–130.
36 Zheng YL, Zhen DK, DeMaria MA, Berry SM, Wapner RJ, Evans MI, Copeland D, Williams JM, Bianchi DW: Search for the optimal fetal cell antibody: Results of immunophenotyping studies using flow cytometry. Hum Genet 1997;100:35–42.
37 Hamada H, Arinami T, Sohda S, Hamaguchi H, Kubo T: Mid-trimester fetal sex determination from maternal peripheral blood by fluorescenece in situ hybridization without enrichment of fetal cells. Prenat Diagn 1995;15:78–81.
38 Slunga Tallberg A, el Rifai W, Keinanen M, Ylinen K, Kurki T, Klinger K, Ylikorbala O: Maternal origin of nucleated erythrocytes in peripheral venous blood of pregnant women. Hum Genet 1995; 96:53–57.
39 Slunga Tallberg A, el Rifai W, Keinanen M, Ylinen K, Kurki T, Klinger K, Ylikorbala O: Maternal origin of transferrin receptor positive cells in venous blood of pregnant women. Clin Genet 1996;49: 196–199.
40 Bianchi DW, Williams JM, Sullivan LM: PCR quantitation of fetal cells in maternal blood in normal and aneuploid pregnancies. Am J Hum Genet 1997;61:822–829.

41 Gänshirt D, Borejesson-Stoll R, Burschyk M, Garritsen HS, Neusser M, Smeets MW, Valesco M, Walde C, Holzgrewe W: Succesful prenatal diagnosis from maternal blood with magnetic-activated cell sorting. Ann NY Acad Sci 1994;731:103–104.
42 Thilaganathan B, Meher-Homji NJ, Nicolaides KH: Blood transferrin receptor expression in chromosomally abnormal fetuses. Prenat Diagn 1995;15:282–284.
43 Andreaux JP, Renard M, Daffos F, Forestier F: Erythropoietic progenitor cells in human fetal blood. Nouv Rev Fr Hématol 1991;33:223–226.
44 Levinson G, Fields RA, Harton GL, Palmer FT, Maddalena A, Fugger EF, Shulman JD: Reliable gender screening for human preimplantantation embryos using multiple dNA target sequences. Hum Reprod 1992;7:1304–1313.
45 Reading JP, Huffman JL, Wu JC, Palmer FT, Harton GL, Sisson ME, Keyvanfar K, Gresinger TH, Cochrane WJ, Fallon LA, Menapace-Drew GF, Cummings EA, Jones SL, Black SH, Schulman JD, Levison G: Nucleated erythrocytes in maternal blood: Quantity and quality of fetal cells in enriched populations. Mol Hum Reprod 1995;10:2510–2515.
46 Geifman-Holtzman O, Holtzman EJ, Vadnais TJ, Phillips VE, Capeless EL, Bianchi DW: Detection of fetal HLA-DQ sequences in maternal blood: A gender-independent technique of fetal cell identification. Prenat Diagn 1995;15:261–268.
47 Pao-Lin K, How-Ran G: Nucleated red blood cells in maternal blood during pregnancy. Obstet Gynecol 1999;94:464–468.
48 Williams JGK, Kubelik AR, Livak KJ, Rafalski JA, Tingey SV. DNA polymorphisms amplified by arbitrary primers are useful as genetic markers. Nucleic Acids Res 1990;18:6531–6535.

Dr. Ilona Hromadníková, 2nd Clinic of Paediatrics, 2nd Medical Faculty, Charles University
University Hospital Motol, V Úvalu 84, CS–150 18 Prague 5 (Czech Republic)
Tel. +420 2 24432258, Fax +420 2 24432220, E-Mail ilona.hromadnikova@lfmotol.cuni.cz

Hahn S, Holzgreve W (eds): Fetal Cells and Fetal DNA in Maternal Blood. New Developments for a New Millennium. 11th Fetal Cell Workshop, Basel, 2000. Basel, Karger, 2001, pp 74–81

Using AMCA to Label the Fetal Cell Antigen in Fetal Erythroblasts Circumvents Heme Autofluorescence, Enhances cFISH Efficiency and Improves Specificity of Fetal Cell Identification

Mahesh Choolani[1,a], *Hilary O'Donnell*[a], *Cesare Campagnoli*[a], *Sailesh Kumar*[a], *Irene Roberts*[b], *Phillip Bennett*[a], *Nicholas M. Fisk*[a]

[a] Department of Maternal and Fetal Medicine, Queen Charlotte's and Chelsea Hospital,
[b] Department of Haematology, Imperial College School of Medicine, Hammersmith Campus, London, UK

Introduction

The enrichment of fetal erythroblasts in maternal blood offers a promising non-invasive alternative for prenatal genetic diagnosis [1]. Obtaining fetal material in this way avoids the risks associated with amniocentesis, chorionic villus sampling and fetal blood sampling [2], and is likely to increase the uptake of prenatal diagnosis for chromosomal and monogenic disorders. The fetal nucleated red blood cell (FNRBC) is the predominant cell type in first-trimester fetal whole blood. It bears the fetal genotype and has a limited lifespan, thereby rendering it unlikely to persist across pregnancies. In addition, it is mononuclear, making it suitable for chromosomal fluorescence in situ hybridisation (cFISH).

Prenatal diagnosis using fetal erythroblasts isolated from maternal blood has been arbitrarily divided into enrichment, identification and diagnostic steps. Currently, fetal erythroblasts are separated using either fluorescence-activated cell sorting (FACS) [3] or magnetically activated cell sorting (MACS) [4] protocols.

[1] M.C. is supported by the National University of Singapore.

Researchers using the former usually combine the sorting process with immunostaining with the γ-globin chain of fetal haemoglobin (HbF). In contrast, those using MACS sort for surface antigens on erythroblasts and subsequently combine identification and diagnosis in situ on a glass slide. The limitation of the first method is that it divorces identification from diagnosis, while that of the second method combining immunocytochemistry (ICC) and cFISH has been the need to switch between light and fluorescence microscopes. Both introduce a potential for error in genetic diagnosis.

Correct identification of fetal cells is the linchpin of a non-invasive prenatal diagnostic technique. The identification system should be fetally specific, gender independent and reliant on developmental differences between mother and fetus. Mavrou et al. [5] suggested that the use of ε-globin as the fetal cell identifier confers greater specificity than γ-globin. We have recently demonstrated that there are sufficient ε-globin-producing fetal erythroblasts in first-trimester fetal blood to make them suitable target cells for non-invasive prenatal diagnosis [Choolani et al., Blood, in press]. We speculate that combining identification and diagnosis such that both can be visualised simultaneously, without needing to register the position of the fetal cell or to switch between light and fluorescence microscopy, will enhance the accuracy of diagnosis.

Autofluorescence of the heme within fetal erythroblasts has limited the development of a co-visualisation technique. Current immunoenzymatic methods combining ICC with cFISH mostly use Vector Blue Substrate® (VBS) as the labelling dye [6]. This light microscopy dye avoids the difficulties in specificity posed by autofluorescence of heme, but necessitates cumbersome switching of microscopes and hinders cFISH efficiency [7]. In this presentation, we demonstrate that labelling the intracellular fetal cell antigen with the blue fluorescent dye AMCA (7-amino-4-methylcoumarin-3-acetic acid) effectively circumvents the problems of autofluorescence and enhances cFISH efficiency.

Materials and Methods

First-trimester fetal whole blood was obtained by ultrasound-guided, transabdominal cardiocentesis immediately prior to clinically indicated surgical termination of pregnancy [8]. The procedure was performed under anaesthesia and under sterile conditions. Blood collection was approved by the institutional ethics committee in compliance with national guidelines regarding the use of fetal tissue for research purposes. All women gave written informed consent.

The nucleated cell count of each sample was determined and 30,000 FNRBCs, suspended in 1% bovine serum albumin in phosphate-buffered saline (PBS), were cytocentrifuged onto each glass slide. Slides were air-dried overnight, used immediately or sealed in aluminium foil and stored at $-20\,^{\circ}C$ until further processing. Stored slides were defrosted at room temperature before being unwrapped.

To determine the effect of autofluorescence, we studied 4 groups of 50 fetal erythroblasts each. All the erythroblasts were from the same sample at 9 weeks of gestation. In 3 groups, the cells were stained using fluorescence immunocytochemistry [9] with either fluorescein isothiocyanate (FITC), phycoerythrin (PE) or AMCA for either the ε- or the γ-globin chain and the fluorescence intensities of positive cells studied. The 4th group was not stained and the autofluorescence within the cells determined through the red (Texas Red®), green (FITC) and blue (DAPI; diamidino-2-phenyl-indole) channels. We used FITC and AMCA to label ε-globin but for PE labelled γ-globin instead because it was commercially available preconjugated. An assumption was made that this difference would not affect the intensity results within individual cells (if at all, the greater concentration of γ-globin anticipated within fetal erythroblasts at this gestation would increase the fluorescence intensity of PE in positive cells).

To compare image intensities captured through the different filters from different areas of different slides, all images were colour normalised (256 grey levels; IPLab Software, Digital Scientific, Cambridge, UK) according to set criteria before analysis. Ten randomly selected clusters of 5 neighbouring cells were studied for each of the 4 study groups. Clusters were labelled 1–10 consecutively upon selection. Within each cell, 10 small areas within the cytoplasm were studied. A mean fluorescence intensity was calculated for each cluster of 5 cells. This number was transformed to a relative fluorescence intensity of cluster (RFIC) by making it a percentage of the 256 grey levels [10]. Within each filter channel, green, red and blue, the difference between the corresponding RFICs was calculated.

A standard XY-FISH protocol for lymphocyte nuclei (Vysis, Downer's Grove, Ill.) was modified and combined with fluorescence immunocytochemistry using AMCA as the ε-globin label within FNRBCs. The modification comprised using methanol and acetone as fixative, permeabilisation with dilute (0.25%) glacial acetic acid and heat denaturation of DNA instead of warm formamide. Slides were analysed by epifluorescence microscopy using a triple band pass filter set for DAPI, FITC and Texas Red®. Images were captured using a cooled charge-coupled device camera and reviewed in Quipps m-FISH software (Vysis). Hybridisation efficiency was determined by scoring 100 consecutive AMCA-positive nucleated cells for the presence of two FISH signals, and compared to the hybridisation efficiency of lymphocytes.

The sensitivity and specificity of fluorescence ICC-cFISH was determined in artificial mixtures of fetal erythroblasts in other cell types. The following artificial mixtures were prepared: maternal nucleated cells, maternal erythrocytes and cord blood laced with male fetal erythroblasts. Slides were also prepared of pure populations of maternal nucleated cells, maternal erythrocytes and cord blood. Results of the fluorescence ICC and combined fluorescence ICC-cFISH experiments were validated using appropriate positive and negative controls.

Non-parametric statistics were derived using SPSS Software (SPSS Inc., Chicago, Ill.). Some of the data presented here have been published elsewhere [Choolani et al., Blood, in press].

Results and Discussion

Any technique that separates the identification step from diagnosis has potential for error, especially if there is a need to switch microscopes [11] or relocate the cell of interest [12]. Heme autofluorescence has limited the development of a technique that allows simultaneous visualisation of the fetal cell identifier and molecular genetic signal. Earlier attempts to reduce or correct for autofluorescence have involved the concomitant use of dyes that absorb emitted wavelengths [13] or mathematical manipulations of pixel intensity [10]. Instead, to circumvent this, we systematically studied autofluorescence within fetal erythroblasts.

We found that the mean difference in the relative fluorescence intensity between stained and unstained erythroblasts was greater with AMCA (mean 43.0, 95% CI 34.2–51.8, SD 13.9) as the reporting label compared with FITC (mean 24.1, 95% CI 16.3–31.9, SD 12.4) or PE (mean 9.8, 95% CI 4.8–14.8, SD 8.0; fig. 1). Viewed through the green channel, 55.9% of the unstained cells have a greater (auto)fluorescence intensity than weakly stained positive cells; this latter group of weakly stained cells represent 23.8% of all positive cells, and 20.1% of all green cells fall within this zone of ambiguity. Similarly, through the red channel, 68.7% of unstained cells have a greater (auto)fluorescence intensity than weakly stained positive cells; this latter group of weakly stained cells represent 58.2% of all positive cells, and 46.0% of all red cells fall within this zone of ambiguity. In the remaining four fifths of cases for green, and half the cases for red, stained and autofluorescent cells can be easily distinguished by optimising the fluorescence threshold on the colour histogram in the image-capture software. There is no overlap of fluorescence between stained and unstained cells viewed through the blue channel suggesting that confusion between positive and negative cells is unlikely. Choosing AMCA as the label for the anti-ε-globin antibody should therefore improve specificity.

In our hands, using standard VBS to label the globin chains appears incompatible with accurate cFISH [6, 7]; hybridisation efficiencies for 2 signals varied widely between 0 and 60%. We have abandoned this strategy and chose instead to compare the combined ICC-cFISH developed in our laboratory with standard cFISH using Carnoy's fixative. Carnoy's (3:1 v/v methanol/glacial acetic acid) is a standard cFISH fixative that generates excellent hybridisation efficiency but severely disrupts the cell membrane and cytoplasm, leaving only the nucleus present and intact on the glass slide. On the other hand, cross-linkers such as formaldehyde or paraformaldehyde, when tested (table 1), resulted in good cell morphology but poor nuclear hybridisation and trapped most of the cFISH probe within the cytoplasm. Methanol appeared necessary to retain good cell morphology and cytoplasmic staining. Increasing exposure to acetone improved the cFISH signal but damaged cell morphology. Leakage of the haemoglobin into the inter-

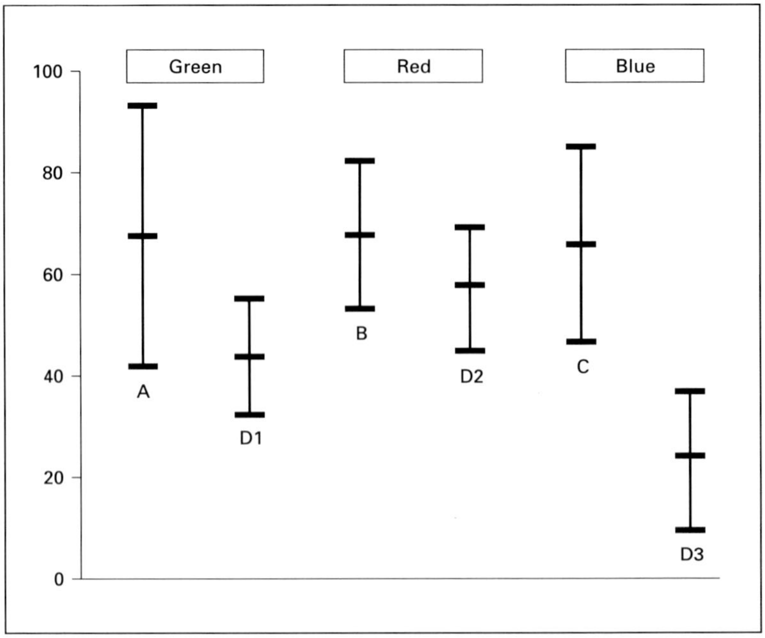

Fig. 1. Relative fluorescence intensity of stained and unstained cells (autofluorescence) through green, red and blue channels. A, B and C represent the mean relative fluorescence intensity and the 95% data intervals (±1.96 SD) of fetal erythroblasts stained with anti-globin antibodies conjugated with FITC, PE and AMCA, respectively. D1, D2, D3 represent the autofluorescence of unstained fetal erythroblasts when viewed through the green, red and blue filters, respectively. There is overlap between the brighter autofluorescent and weakly stained cells in the green and red channels but not in the blue.

cellular spaces between the erythroblasts was inversely related to the retention of cell morphology and integrity of the cell membrane. Semi-qualitative experiments revealed that 1:1 v/v methanol:acetone was the optimal fixative that was faithful to cell morphology and cell membrane integrity. Permeabilisation with 0.25% glacial acetic acid further improved cFISH efficiency to 97%, comparable to the 98% ($z = 0.74$; NS) obtained in control slides of lymphocyte nuclei using the standard cFISH protocol.

Since the technique was designed to detect rare events, it was necessary to determine the limit of sensitivity of the technique and its specificity under rare-event conditions: the rarer the event, the greater the impact of inaccurate diagnosis. None of the maternal nucleated cells, maternal erythrocytes or neonatal erythrocytes stained positively for ε-globin. Within artificial mixtures, fluorescence

Table 1. Comparison of fixatives for combined fluorescence ICC and cFISH

Fixative	Cell Morphology	Haemoglobin	FISH Efficiency
Methanol:glacial acetic acid			
3:1	Cell completely disrupted	Completely lost	Excellent
4:1	Cell completely disrupted	Completely lost	Excellent
9:1	Cell completely disrupted	Completely lost	Excellent
19:1	Poor	Some Hb intact	Good
Paraformaldehyde (2 min)			
4%	Excellent	Completely intact	Poor
Formaldehyde (2 min)			
2%	Excellent	Completely intact	Poor
4%	Excellent	Completely intact	Poor
5%	Excellent	Completely intact	Poor
Methanol:acetone (2 min)			
1:0	Excellent	Completely intact	Poor
19:1	Excellent	Completely intact	Poor
15:1	Excellent	Completely intact	Poor
10:1	Excellent	Completely intact	Poor
5:1	Excellent	Completely intact	Poor
3:1	Excellent	Completely intact	Fair
1:1	*Excellent*	*Completely intact*	*Fair*
1:3	Good	Completely intact	Good
1:5	Good	Completely intact	Good
1:10	Good	Most Hb intact	Good
1:15	Poor	Some Hb intact	Excellent
1:19	Poor	Most Hb lost	Excellent
0:1	Poor	Most Hb lost	Excellent

Use of acetic acid disrupts cell membrane integrity and cell morphology. Cross-linking fixatives give good cell morphology but poor cFISH efficiency. Methanol is faithful to cell integrity and methanol:acetone (1:1) allows the best morphology and haemoglobin retention with reasonable cFISH probe penetration and hybridisation.

ICC-cFISH was sensitive enough to detect one ε-positive fetal erythroblast among 10^5 nucleated cells or 10^6 erythrocytes ($z = 1.99$; $p < 0.05$). The reason for this is unclear, but may reflect the method of preparation of the mixture, rather than real differences in sensitivity. The detection of positive cells was easy: positive cells were bright blue within a black background. In these artificial mixtures, specificity was 100% but the ability of the technique to identify fetal erythroblasts from an artificial mixture containing maternal NRBCs needs to be evaluated.

Two unanticipated observations were made which appear to facilitate this technique. First, heme-containing erythroblasts autofluoresce through the SpectrumAqua® channel as well, whereas the other nucleated cells in peripheral blood do not. This unique difference can be exploited during erythroblast sorting by FACS or their identification as heme-containing nucleated cells in situ. Second, AMCA had a tendency to accumulate around the nucleus. This unexplained but fortuitous occurrence negates the need for nuclear counterstaining during cFISH, because AMCA acts as surrogate counterstain.

The ability to visualise concurrently a fetal cell marker and a hybridisation signal within the interphase nucleus represents a significant advance towards the goal of prenatal diagnosis using fetal cells derived from maternal blood. Using AMCA to label the intracellular fetal antigen circumvents the problems associated with autofluorescence, enhances cFISH hybridisation efficiency and should improve the specificity of diagnosis by permitting simultaneous visualisation of the fetal cell identifier and nuclear FISH. Direct labelling with AMCA-conjugated anti-ε antibody should further abbreviate the procedure time (circa 8 h), bringing this technology closer to clinical practice. Although the diagnostic step was developed for cFISH, this technique could also be adapted to in situ PCR to diagnose monogenic disorders.

References

1 Bianchi DW: Fetal cells in the maternal circulation: Feasibility for prenatal diagnosis. Br J Haematol 1999;105:574–583.

2 Sebire NJ, von Kaisenberg C, Nicolaides KH: Diagnostic techniques; in Snijders RJM, Nicolaides KH (eds): Ultrasound Markers for Fetal Chromosomal Defects. London, Parthenon, 1996, pp 157–170.

3 Herzenberg LA, Bianchi DW, Schroder J, et al: Fetal cells in the blood of pregnant women: Detection and enrichment by fluorescence-activated cell sorting. Proc Natl Acad Sci USA 1979;76:1453–1455.

4 Gänshirt-Ahlert D, Burschyk M, Garritsen HS, et al: Magnetic cell sorting and the transferrin receptor as potential means of prenatal diagnosis from maternal blood. Am J Obstet Gynecol 1992;166:1350–1355.

5 Mavrou A, Kolialexi A, Zheng YL, et al: Improved specificity of NRBC detection in chorionic villus sample supernatant fluids using anti-zeta and anti-epsilon monoclonal antibodies. Fetal Diagn Ther 1999;14:291–295.

6 Oosterwijk JC, Mesker WE, Oowerkerk-van Velzen MC, et al: Development of a preparation and staining method for fetal erythroblasts in maternal blood: Simultaneous immunocytochemical staining and FISH analysis. Cytometry 1998;32:170–177.

7 Wang JY, Zhen DK, Falco VM, et al: Fetal nucleated erythrocyte recovery: Fluorescence activated cell sorting-based positive selection using anti-gamma globin versus magnetic activated cell sorting using anti-CD45 depletion and anti-gamma globin positive selection. Cytometry 2000;39:224–230.

8 Campagnoli C, Fisk N, Overton T, et al: Circulating hematopoietic progenitor cells in first trimester fetal blood. Blood 2000:95:1967–1972.

9 Mesker WE, Ouwerkerk-van Velzen MC, Oosterwijk JC, et al: Two-colour immunocytochemical staining of gamma and epsilon type haemoglobin in fetal red cells. Prenat Diagn 1998;18:1131–1137.
10 Szollosi J, Lockett SJ, Balazs M, Waldman FM: Autofluorescence correction for fluorescence in situ hybridization. Cytometry 1995;20:356–361.
11 Pazouki S, Hume R, Burchell A: A rapid combined immunocytochemical and fluorescence in situ hybridisation method for the identification of human fetal nucleated red blood cells. Acta Histochem 1996;98:29–37.
12 Sekizawa A, Taguchi A Watanabe A, et al: Analysis of HLA-DQ alpha sequences for prenatal diagnosis in single fetal cells from maternal blood. Hum Genet 1998;102:393–396.
13 Mosiman VL, Patterson BK, Canterero L, Goolsby CL: Reducing cellular autofluorescence in flow cytometry: An in situ method. Cytometry 1997;30:151–156.

Dr. Mahesh Choolani, Institute of Reproductive and Developmental Biology
4th Floor, Imperial College School of Medicine
Hammersmith Campus, Du Cane Road, London W12 0NN (UK)
Tel. +44 20 7594 2121, Fax +44 20 7594 2129, E-Mail mchoolani@cwcom.net

Hahn S, Holzgreve W (eds): Fetal Cells and Fetal DNA in Maternal Blood. New Developments for a New Millennium. 11th Fetal Cell Workshop, Basel, 2000. Basel, Karger, 2001, pp 82–90

Fetal Cell Detection in Maternal Blood Using the Solid Phase Cytometer Chem*Scan*®RDI: Preliminary Studies

Laurence Serradell[a], *Jean-Michel Dupont*[b], *Françoise Ferré*[c], *Didier Rabineau*[b], *Philippe Cornet*[a]

[a] Chemunex, Ivry-sur-Seine;
[b] Service d'Histologie-Embryologie-Cytogénétique, Hôpital Cochin, Paris, and
[c] U361 INSERM, Pavillon Baudelocque, Paris, France

Introduction

Today the prenatal diagnosis of chromosomal abnormalities is performed by fetal karyotyping using invasive samplings involving amniocentesis, choriocentesis or cordocentesis. It is well known that these methods present important risks to the fetus and the mother and cannot be used for routine analysis in pregnant women.

Moreover, the current biochemical markers for detecting any chromosomal abnormalities are not reliable enough as 21% of babies harboring trisomy 21 are not detected. It is obvious that new methods are needed to overcome the limitations of the current prenatal diagnostic methods.

One promising and exciting approach would be the direct detection and analysis of fetal cells from maternal blood [1, 2]. Because fetal cells are found at very low frequencies in maternal blood ($1/10^6$–$1/10^7$ nucleated cells), an automated system to detect rare cellular events would be an important advantage for this purpose and numerous other biomedical applications [3].

This article reports on a preliminary feasibility study for labelling and detecting fetal cells in maternal blood with an ultrasensitive analyzer: the Chem*Scan*® RDI system.

Patients and Methods

Sample and Cell Selection

Peripheral venous blood samples (10 ml) were obtained with informed consent from 12 pregnant women (8–18 weeks of gestation).

Cytotrophoblast cells were isolated and purified from placentas (8–12 weeks of gestation). The isolation procedure involves enzymatic digestion of placenta following Percoll® centrifugation. Enrichment was completed by negative selection using magnetic beads coated with anti-CD45.

Peripheral blood mononuclear cells (PBMCs) were prepared from the blood of healthy donors after Ficoll® centrifugation.

The Comité Consultatif de Protection des Personnes dans la Recherche Biomédicale, Paris-Cochin, was in agreement with this study.

Immunofluorescent Staining

Antibodies. Trophoblast targeting was performed with GB25 (Prof. B. Rossi, INSERM U364, Nice), and anti-CK7 (Dako) which recognizes cytokeratine 7. Erythroblast targeting was performed with anti-γ-fetal hemoglobin (Europa Products); anti-ε-fetal hemoglobin (Europa Bioproducts), and anti-CD71 (Santa Cruz Biotechnology Inc.).

Staining Protocol. The fluorochrome used was Fast Red combined with the kit Vector® Red (Vector) which gave good fluorescent signals on positive cells and was compatible with the Chem*Scan*®RDI analyzer. Silanized glass slides and silanized γ-maleimidobutyrique acid-treated slides were used for all the experiments.

Fluorescence in situ Hybridization (FISH) Staining

FISH probes were supplied by Vysis™. Two kinds of probe mixes were used: (1) centromeric probes X and Y labelled in green (Spectrum Green) and blue (Aqua), respectively, and (2) the Aneuvysion kit which is composed of 5 probes specific for chromosomes X, Y 18, 13, and 21.

FISH Protocol

After the immunolabelling step, the slides were processed to perform a second fixation step followed by dehydration, hybridization and washing steps.

Slide Analysis

The slides were analyzed by the Chem*Scan*®RDI system which is a laser-scanning device manufactured by Chemunex (Ivry-sur-Seine, France). This new biological analyzer based on solid-phase cytometry has been described elsewhere [4] and has been proved to very rapidly detect rare fluorescent events when scanning an area of 6 cm^2 in 3 min.

The main features of this system, namely the laser scanning concept, the fluorescent signal discrimination and the automatic microscope visualization, are summarized in figures 1–3, respectively.

The cells detected by the Chem*Scan*®RDI were directly observed with a Leica microscope equipped with an epifluorescence system.

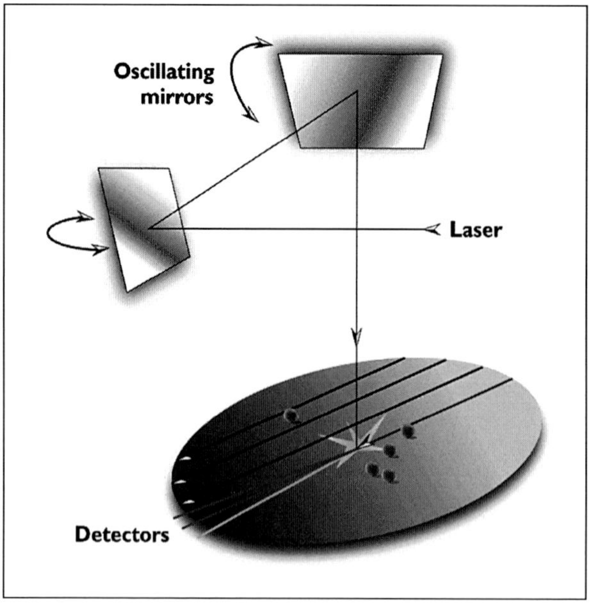

Fig. 1. Chem*Scan*®RDI laser scanning concept.

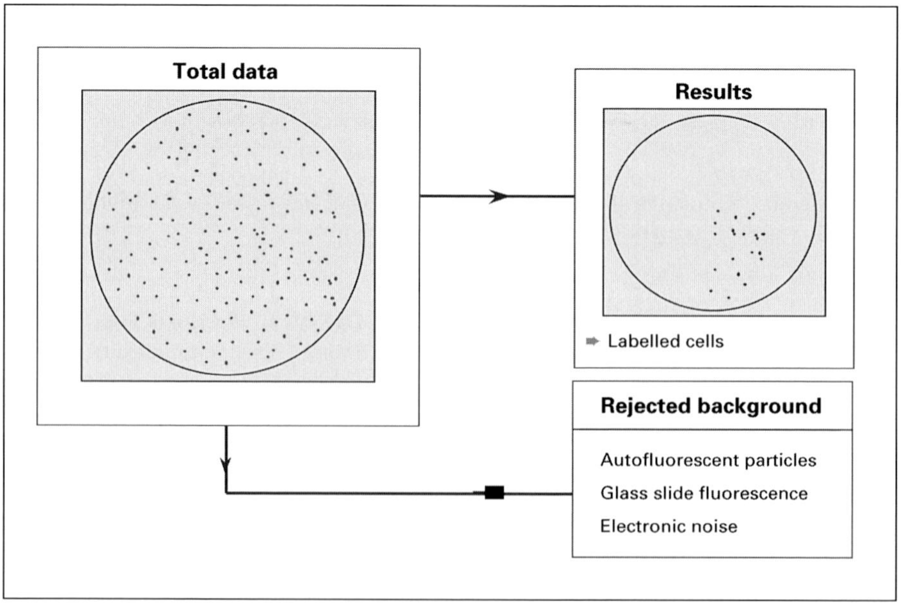

Fig. 2. Chem*Scan*®RDI signal discrimination.

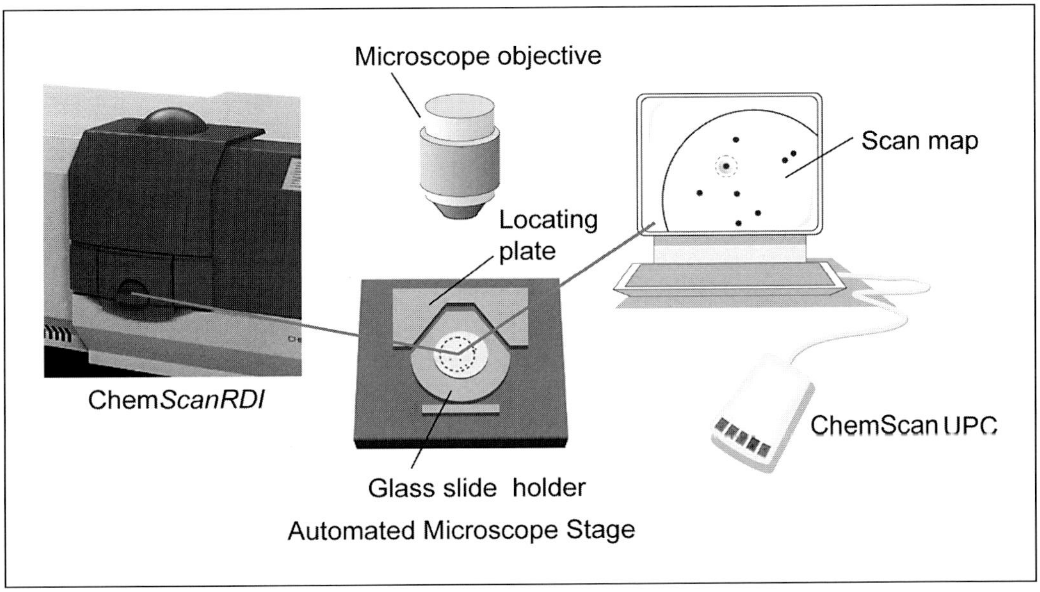

Fig. 3. Automatic microscope visualization.

Results

Feasibility Study of the Cellular Model Protocol

The immuno-FISH protocol combined with the Chem*Scan*®RDI system for detection of the cells of interest was assessed on a cellular model depicted in figure 4. Slides containing PBMCs from a male donor and a few cytotrophoblasts purified from a female placenta harboring trisomy 18 were submitted to the immuno-FISH protocol before being analyzed with the Chem*Scan*®RDI. Only the positive cells (trophoblasts) labelled in red after the immunolabelling step were detected with the Chem*Scan*®RDI.

When observed with the epifluorescence microscope for chromosome content, rare cells show two green spots (XX) and three blue spots (3 × 18) as expected, whereas the nonlabelled cells display one green spot (X), one orange spot (Y) and two blue spots (2 × 18) confirming their PBMC origin (fig. 5). It was verified that all the positive cells were effectively detected by the Chem*Scan*®RDI after manually rescanning the slides (data not shown).

From these first observations, it can be emphasized that the selected mix of antibodies combined with the immuno-FISH protocol and the Chem*Scan*®RDI analysis allows the detection and selection of specifically purified trophoblasts,

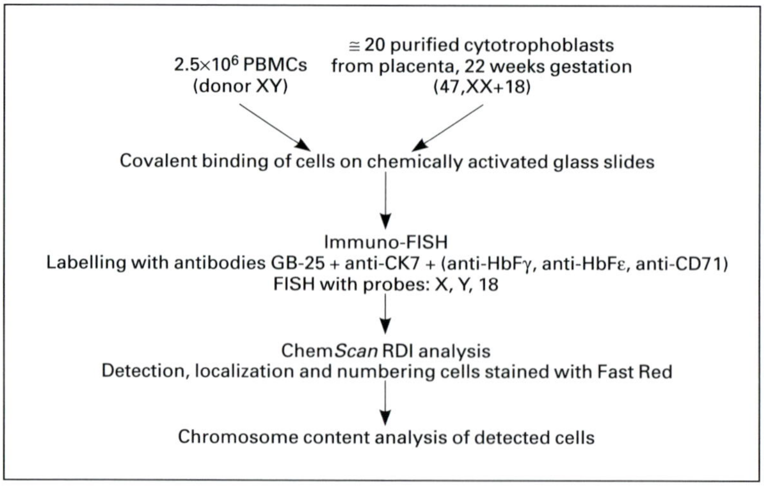

Fig. 4. Application of the immuno-FISH protocol on a cellular model.

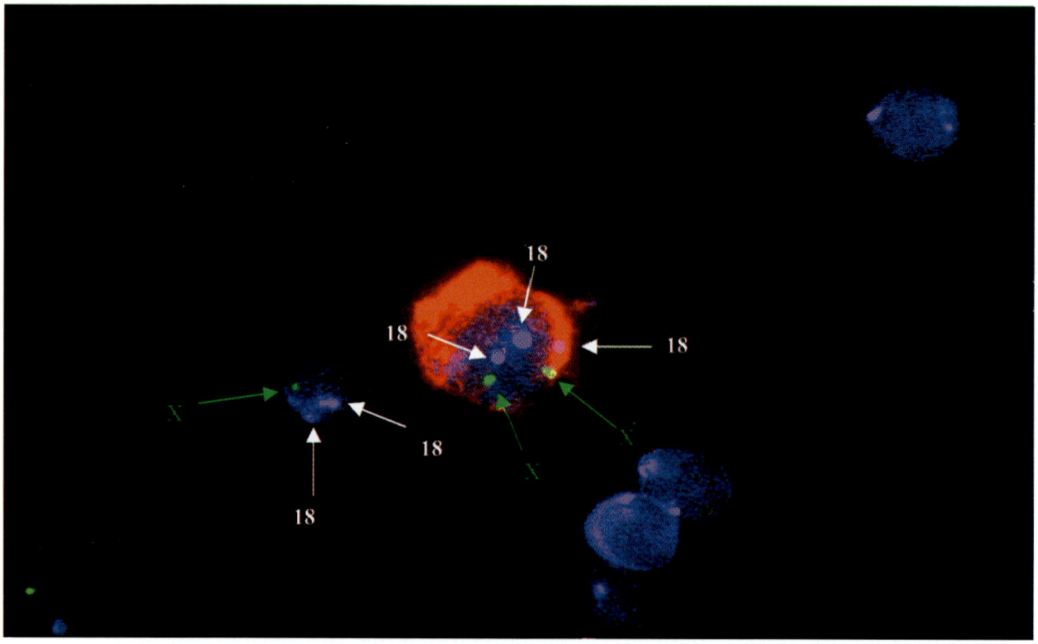

Fig. 5. Purified trophoblast cell (47,XX+18) in PBMCs (XY) by immuno-FISH labelling.

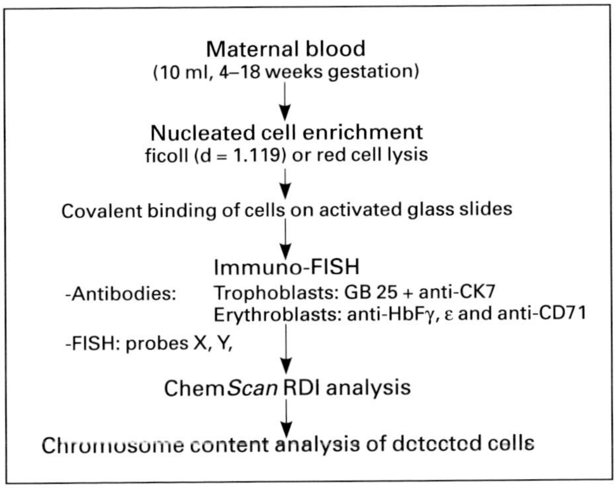

Fig. 6. Application of the immuno-FISH protocol for the analysis of peripheral blood from pregnant women.

the reliable identification of sexual chromosomes, and underline an aneuploidy (trisomy 18).

*Application of Immuno-FISH Protocol and Chem*Scan®*RDI Analysis for Peripheral Blood Analysis from Pregnant Women*

The protocol, which has been developed and successfully assessed as mentioned above on an artificial cellular model, has been extrapolated and evaluated on real samples, i.e. the blood of pregnant women, according to the procedure described in figure 6. Under these conditions the number of cells deposited on the slides is about 1.5×10^6 cells. The preliminary results obtained from 12 patients are summarized in table 1.

The patients for whom analyses have been performed can be divided in 2 groups. (1) Patients carrying a male fetus, in whose blood no Fast Red-positive cells were detected although in some of them (CFC 2, 13, 15, FP) a significantly high number of nucleated cells were analyzed. These negative results cannot be explained by a problem with the immuno-FISH protocol as systematic control of the slides containing a mixture of male donor PBMCs and some cytotrophoblasts of karyotype (47,XX+18) were processed at the same time and in same way as the patient samples. (2) Patients in whose blood some positive cells (red cells) were detected by the Chem*Scan*®RDI. All these positive cells had two XX spots (fig. 7). These positive cells of female karyotype were detected in patients carrying a female fetus (CFC 6, 16 and RM) and can be considered as real fetal cells or can

Table 1. Analysis of peripheral blood from pregnant women using the immuno-FISH protocol

Patient	Age years	Gestation weeks	Fetal genotype	Number of nucleated cells (1.5 × 10^6/slide)	Labeled cells detected	
					Fast Red	FISH
CFC 15	42	10	46,XY	17	0	
BA	38	11	46,XY	4	0	
CFC 13	44	10	46,XY	25	0	
CFC 2	39	13	46,XY	12	0	
CFC 12	40	8	46,XY	2.6	0	
FP	40	9	–	28	0	
CFC 16	41	9	46,XX	10	10	XX
RM	44	5	–	26	14	XX
CFC 6	39	15	46,XX	8	13	XX
CFC 14	42	4	46,XY	5.8	3	XX
HI	40	13	46,XY	26	9	XX
ED	39	18	46,XY	10.9	6	XX

have a maternal origin, in this case they would be false positives. At the present time, the formal origin of these cells remains an open question.

Some XX red cells have been detected in patients carrying a male fetus (CFC 14, HI and ED) and have to be considered unambiguously as true false positives.

Discussion

Fetal cells circulating in maternal blood would be an excellent source for a noninvasive prenatal diagnosis procedure without risks to the fetus or the mother. Since fetal cells are known to be rare, with a frequency of about 10^{-6} or less, their detection and characterization are a true challenge. The strategy we applied was to simultaneously probe the trophoblast and erythroblast cells and to perform a single and simple enrichment step of the nucleated blood cells by chemical lysis of the red blood cells or by density gradient.

Fetal cells are characterized by an innovative immunolabelling step combined with a FISH procedure. The rare labelled positive cells are detected and located on the surface of a glass slide thanks to a new and efficient slide analyzer, the Chem*Scan*®RDI perfectly adapted for the detection of rare events.

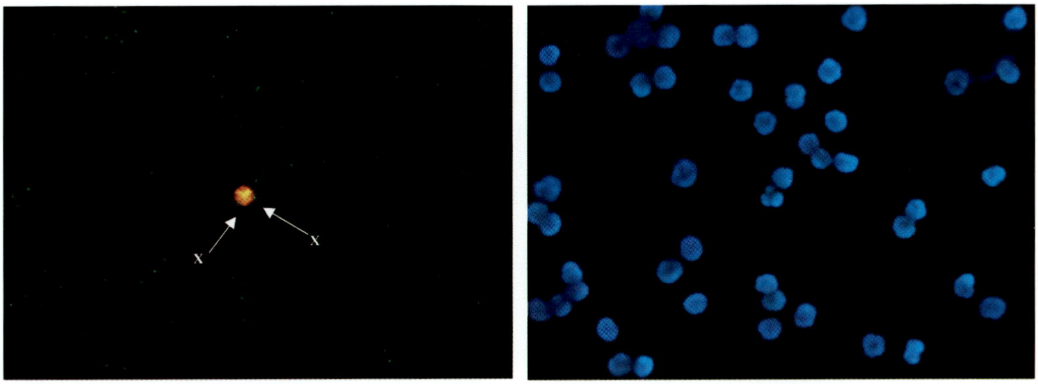

Immuno-FISH Hoechst

Fig. 7. Chem*Scan*®RDI detected immuno-FISH-labelled cells from the blood of a pregnant woman.

The first part of this study has shown that by using an experimental cellular model, an efficient labelling procedure of the cells of interest was able to give good fluorescent signals and was compatible with the visualization of FISH signals and a good detection with the Chem*Scan*®RDI. This study has demonstrated that with the Chem*Scan*®RDI it was possible (1) to specifically label, detect and select a few cytotrophoblasts among an important population of nucleated cells of peripheral blood origin, and (2) to identify the chromosome content based on the presence of sexual chromosomes and also to identify an aneuploidy (trisomy 18). The immuno-FISH protocol, developed in an experimental model, has been applied to maternal blood samples.

So far, the results obtained cannot formally assert that fetal cells have been detected in maternal blood. Only the detection of XY red cells in the blood of pregnant women carrying a male fetus would constitute irrefutable evidence of their fetal origin.

Some hypotheses can been raised to explain these observations. (1) The immuno-FISH protocol previously developed from a cellular model is not directly applicable to the circulating fetal cells in maternal blood. These latter cells could have a different phenotype than the purified cytotrophoblast cells and could not be recognized by the chosen mix of antibodies. (2) There is an important loss of fetal cells during the different steps of the blood sampling and labelling protocol.

Preliminary experiments have shown that whatever the type of the enrichment phase, up to 90% cellular loss could be noticed from mixing experiments (donor blood spiked with purified cytotrophoblast cells in different ratios) after performing the immuno-FISH labelling protocol.

It can be concluded that the initial low number of fetal cells present in the blood samples combined with a significant cellular loss during the different steps of the protocol are probably at the origin of the absence of formally identified fetal cells in the blood samples analyzed.

References

1 Hahn S, Sant R, Holzgreve W: Fetal cells in maternal blood: Current and future perspectives. Mol Hum Reprod 1998;4:101–107.
2 Holzgreve W: Will ultrasound screening and ultrasound-guided procedures be replaced by non-invasive techniques for the diagnosis of fetal anomalies? Ultrasound Obstet Gynecol 1997;9:217–219.
3 Hamada H, Arinami T, Kubo T, Hamguchi H, Iwasaki H: Fetal nucleated cells in maternal peripheral blood: Frequency and relationship to gestational age. Hum Genet 1993;91:427–432.
4 Mignon-Godefroy K, Guillet J-G, Butor C: Solid phase cytometry for detection of rare events. Cytometry 1997;27:336–344.

Dr. Philippe Cornet, Chemunex, allée de la Seine, F–94854 Ivry-sur-Seine (France)
Tel. +33 1 49 59 20 00, Fax +33 1 49 59 20 01, E-Mail cornet@worldnet.fr

Hahn S, Holzgreve W (eds): Fetal Cells and Fetal DNA in Maternal Blood. New Developments for a New Millennium. 11th Fetal Cell Workshop, Basel, 2000. Basel, Karger, 2001, pp 91–96

Fetal CD34+ Cell Culture and Identification from Maternal Blood

G. Coata[a], *F. Tilesi*[a], *A. Tabilio*[b], *G.C. Di Renzo*[a]

[a] Center of Perinatal and Reproductive Medicine and
[b] Department of Clinical and Experimental Medicine, University of Perugia, Italy

Introduction

In clinical practice, noninvasive prenatal diagnosis has one major limitation. This is that only a few cells can be sorted from maternal blood samples [1, 2] and those which can be are insufficient for a full genetic diagnosis. The main reason for failure is that a specific fetal cell marker which can distinguish and sort fetal cells from maternal blood has not been identified [3, 4]. One strategy for overcoming this obstacle could be to culture those fetal cells which proliferate in vitro [5, 6] such as the CD34+ stem cells which can be cloned [7].

The present study was performed in an attempt to expand CD34+ fetal stem cells in vitro using our combined methodology of enrichment and analysis of CD34+ cell culture by the fluorescence in situ hybridization (FISH) method for a noninvasive prenatal diagnosis of genetic abnormalities.

Materials and Methods

Patients

After informed consent, 20-ml peripheral maternal blood samples were obtained in heparinized tubes from 8 primigravidae at 11–16 weeks of gestation. All were carrying a male fetus (as visualized by ultrasound). Blood samples were taken before amniocentesis. Karyotyping or birth confirmed the fetal sex. Four primigravidae at 11–16 weeks of gestation with female fetuses were also analyzed.

As controls, peripheral blood samples from 4 never pregnant adult females and 3 adult males were analyzed. None of the pregnant women and female controls had a history of blood transfusion or abortion before blood sampling.

Enrichment of Hematopoietic Progenitor Cells (CD34+ Stem Cells)

In order to obtain a pure CD34+ stem cell population from the peripheral blood sample, mononuclear blood cells (peripheral blood mononuclear cells, PBMCs) were isolated by Ficoll-Paque density separation (1.077 g/ml density gradient, Sigma Diagnostic, St. Louis, Mo., USA). Isolated and washed PBMCs were split into two equal parts: one was cultured in a semisolid methylcellulose medium and the other underwent CD34+ stem cell enrichment.

CD34+ stem cells were enriched using MiniMacs magnetically activated cell sorting (Miltenyi Biotech, Bergisch Gladbach, Germany) as reported previously [7]. One part of the enriched CD34+ cells was seeded in a semisolid methylcellulose culture medium and the other part was frozen at –20 °C until further use.

Cell Cultures

1×10^5 non-enriched PBMCs and 1×10^4 CD34+ stem cells were cultured in a Iscove MDM medium (MethoCult GF H4435, StemCell Technologies, Vancouver, B.C., Canada) and incubated at 37 °C in humidified 5% CO_2 for 14 days as reported previously [7].

Colonies were then observed under an inverted microscope and were counted and classified as BFU-E/CFU-E (burst-forming unit-erythroid/colony-forming unit-erythroid) and CFU-GM (colony-forming unit-granulocyte macrophage).

Colony cells were collectively harvested, counted using a Burker hemocytometer and expressed as total number $\times 10^5$ cells [7].

Preparation of Slides for FISH

After culture, CD34+ cells were treated with hypotonic KCl solution (0.56%) and incubated for 30 min in a 37 °C water bath. Cells were fixed with Carnoy solution (3:1 methanol:glacial acetic acid) and incubated for 30 min at 4 °C. After centrifugation at 1,200 rpm for 10 min at room temperature, cells were gently resuspended in 400 μl of fresh Carnoy solution. Two or three drops of cell solution (each drop was 15–25 μl of the cell suspension) were seeded directly on a glass slide placed on the rack in a water bath at 67 °C for at least 10 min. The slides were frozen at –20 °C and stored until use.

Fluorescent Probes

To distinguish male fetal cells from the maternal, X and Y chromosomal analysis was performed using CEP X (α satellite) Spectrum Green hybridizing centromere of human chromosome X (bands Xp 11.1-q11.1, locus DXZ1) and CEP Y (α satellite) Spectrum Orange hybridizing centromere of human chromosome Y (bands Yp11.1, locus DYZ3; Vysis, Downers Grove, Ill., USA).

FISH Analysis

Using FISH we analyzed samples from nucleated cells (obtained after enrichment and culture) from all normal pregnancies. In addition, in 6 of 8 blood samples from pregnant women with male fetuses, we performed FISH in samples obtained after culture of non-enriched PBMCs and enriched CD34+ cells in order to evaluate the efficacy of CD34+ enrichment and recovery of fetal stem cells and potential clonogenicity.

Hybridization was performed according to the manufacturer's instructions. The probe and slides were incubated in a denaturation solution (70% formamide in 2× sodium chloride/sodium citrate (SSC), pH 7.3) in a water bath, for 5 min at 73 °C. Once denatured, slides were dehydrated through an ethanol series (70, 85, and 100%) and warmed in a 42 °C

Table 1. Gestational age, timing of maternal blood sample, fetal karyotypes and cells undergoing FISH analysis in 12 pregnant women

Patient No.	Gestational age, weeks	Timing of sample	Fetal karyotype
1	20	No invasive procedure	46,XY
2	14	Pre-amnio	46,XY
3	19	No invasive procedure	46,XY
4	16	Pre-amnio	46,XY
5	16	Pre-amnio	46,XY
6	14	Pre-amnio	46,XY
7	16	Pre-amnio	46,XY
8	12	No invasive procedure	46,XY
9	19	No invasive procedure	46,XX
10	16	Pre-amnio	46,XX
11	16	Pre-amnio	46,XX
12	14	Pre-amnio	46,XX

FISH analysis of cultured CD34+ nucleated cells in normal pregnancies.

oven. Ten microliters of probe mixture were applied. Slides were then sealed using glass coverslips and rubber cement. Hybridization was carried out at 42 °C overnight. After hybridization, the coverslips were removed and slides were washed 3 times for 10 min in a post-wash solution (50% formamide/2 × SSC), followed by one 10-min wash in 2 × SSC (pH 7.3) and one 5-min wash in 2 × SSC/1% NP-40. All washes were performed at 46 °C. After the last wash, the slides were counterstained with 10 µl of DAPI II (Vysis, Downers Grove, Ill., USA) and sealed using coverslips and nail polish.

Slides were analyzed by a fluorescence microscopy (Olympus BH-2, Japan) equipped with a triple bandpass filter set (DAPI/Green/Orange, Vysis, Downers Grove, Ill., USA) designed to visualize of Spectrum Green and Spectrum Orange. The fluorescence microscope is equipped with a PM20 automatic photomicrographic system (Olympus, Japan).

Male (fetal) cells had one red and one green fluorescent signal; female (maternal) cells had two green fluorescent signals. The fetal cells were scored and the percentage was calculated.

Results

Table 1 reports gestational age, timing of blood samples, fetal karyotypes and cells undergoing FISH analysis in 12 blood samples from pregnant women.

Table 2 shows the results of FISH analysis on enriched, cultured CD34+ cells in normal pregnancy. The range of visualized fetal cells in pregnancies with male

Table 2. FISH analysis of nucleated cells obtained after culturing of enriched CD34+ cells from pregnant women with normal fetuses

Patient No.	After enrichment of CD34+ and culture			
	examined cells	cells with XY signals	purity of fetal cells, %	fetal/maternal cell ratio
1	1,433	6	0.41	1/239
2	1,459	4	0.27	1/364
3	564	6	1.06	1/94
4	570	11	1.90	1/51
5	335	3	0.89	1/110
6	2,499	3	0.12	1/833
7	100	4	4	1/25
8	394	3	0.76	1/131
9	669	0	0	0
10	597	0	0	0
11	701	0	0	0
12	523	0	0	0

Patients 1–8 were pregnant with male fetuses, and patients 9–12 were pregnant with female fetuses.

Table 3. FISH analysis of nucleated cells after PBMC culture and after CD34+ enrichment

Patient No.	After PBMC culture			After CD34+ enrichment		
	examined cells	scored fetal cells	fetal/maternal cell ratio	examined cells	scored fetal cells	fetal/maternal cell ratio
1	1,283	1	1/1,283	1,433	6	1/239
2	780	0	0	1,459	4	1/364
3	560	1	1/560	564	6	1/94
4	747	5	1/149	570	11	1/51
5	669	5	1/133	335	3	1/110
8	421	1	1/421	394	3	1/131

fetuses was from 3 to 11 fetal cells in a range of scored maternal cells from 100 to 2,499 cells. In the 4 pregnancies with female fetuses, we did not observe any cell with XY fluorescent signals in cultured cells.

Table 3 shows the efficacy of CD34+ enrichment in pregnancies with normal fetuses, as determined by the results of FISH analysis in 6 of 8 pregnant women studied with male fetuses. After enrichment and culture of CD34+ cells, we found that the number of XY fetal cells detected in maternal blood increased from 0 to 5 and the fetal/maternal ratio increased from 0 to 1/560 after culture of PBMC, and after CD34+ cell enrichment they increased from 4 to 11 and the fetal/maternal cell ratio from 1/51 to 1/364 after CD34+ cell enrichment. Interestingly, in sample 2 we did not observe fetal XY cells after PBMC culture, but detected male fetal cells after enrichment and culture of CD34+ cells.

Discussion

In the present study, we combined the two methodologies of enrichment of CD34+ hematopoietic cells, by magnetic cell sorting and proliferation of these cells in a semisolid methylcellulose medium to obtain fetal clones. Finally using FISH with an XY-chromosome-specific DNA probe, we analyzed the nucleated cells obtained after 'in vitro' expansion for the purpose of noninvasive prenatal diagnosis.

After enrichment and culture of CD34+ hematopoietic cells, we observe an elevated number of fetal cells in regard to maternal cells with an overall fetal cell purity percentage ranging from 0.12 to 4%. Moreover when we evaluated the efficacy of the enrichment of CD34+ cells, as determined by the recovery of fetal stem cells, we found that they had significantly increased in the enriched CD34+ fraction with regard to the non-enriched PBMC fraction. When the nucleated cells obtained after PBMC culture (without enrichment of CD34+ cells) were analyzed by FISH, we found that the number, the ratio of fetal/maternal cells and the percentage of sex prediction were lower with regard to the results obtained after the enrichment and the culture of CD34+ cells.

In conclusion, we have demonstrated that our strategies permit a slight but promising increase in the number of fetal cells after appropriate enrichment and culture of hematopoietic cells sorted from the maternal circulation. Interestingly in 2 samples, we observed 2 fetal metaphases. This observation proves definitively that fetal cells can be divided actively in vitro, and consequently it is possible to improve the number of fetal cells isolated from the maternal circulation, obtaining a sufficient quantity of fetal DNA for genetic analysis or prediction of fetal sex.

References

1 Price JO, Elias S, Wachtel SS, Klinger K, Dockter M, Tharapel A, Shulman LP, Phillips OP, Meyers CM, Shook D, Simpson JL: Prenatal diagnosis with fetal cells isolated from maternal blood by multiparameter flow cytometry. Am J Obstet Gynecol 1991;165:1731–1736.

2 Hamada H, Arinami T, Kubo T, Hamaguchi H, Iwasaki H: Fetal nucleated cells in maternal peripheral blood: Frequency and relationship to gestational age. Hum Genet 1993;91:427–432.

3 Bianchi DW, Klinger KW, Vadnais TJ, DeMaria MA, Shuber AP, Skoletsky J, Midura P, Dirisio M, Pelletiers C, Genova M, Erikson MS, Williams JM: Development of a model system to compare cell separation methods for the isolation of fetal cells from maternal blood. Prenat Diagn 1996;16: 289–298.

4 Zheng YL, Zhen DK, DeMaria MA, Berry SM, Wapner RJ, Evans MI, Copeland D, Williams JM, Bianchi DW: Search for the optimal fetal cell antibody: Results of immunophenotyping studies using flow cytometry. Hum Genet 1997;100:35–42.

5 Little MT, Langlois S, Douglas R, Lansdorp PM: Frequency of fetal cells in sorted subpopulations of nucleated erythroid and CD34+ hematopoietic progenitor cells from maternal peripheral blood. Blood 1997;89:2347–2358.

6 Valerio D, Aiello R, Altieri V, Malato AP, Fortunato A, Canazio A: Culture of fetal erythroid progenitor cells from maternal blood for a non-invasive prenatal genetic diagnosis. Prenat Diagn 1996;16:1073–1082.

7 Tilesi F, Coata G, Pennacchi L, Lauro V, Tabilio A, Di Renzo GC: A new methodology of fetal stem cell isolation, purification and expansion: Preliminary results for non invasive prenatal diagnosis. J Hematother Stem Cell Res 2000;9:583–590.

Prof. Gian Carlo Di Renzo, Center of Perinatal and Reproductive Medicine
Policlinico Monteluce, Via Brunamonti, I–06122 Perugia (Italy)
Tel. +39 075 5720563, Fax +39 075 5729271, E-Mail direnzo@unipg.it

Hahn S, Holzgreve W (eds): Fetal Cells and Fetal DNA in Maternal Blood. New Developments for a New Millennium. 11th Fetal Cell Workshop, Basel, 2000. Basel, Karger, 2001, pp 97–103

Free Fetal DNA in Maternal Plasma and Serum: Diagnostic and Clinical Implications

Sinuhe Hahn, Xiao Yan Zhong, Martin R. Bürk, Wolfgang Holzgreve

Department of Obstetrics and Gynecology, University of Basel, Switzerland

Identification of Multiple Fetal Loci by Multiplex PCR

Inspired by reports by Lo et al. [1, 2] on the successful identification of Y chromosome-containing sequences, as well as the determination of fetal rhesus D status, we decided to test the multiplex PCR assays we had developed for the analysis of these two fetal loci on single cells [3] and on free fetal DNA isolated from maternal plasma. We have decided to focus on these two fetal genes, on the one hand, due to their clinical relevance in that there are close to 200 different X-linked disorders characterized to date and since invasive procedures to rescue babies affected by hemolytic disease can still make headlines, and also due to their relative ease in that both genes are absent from the maternal genome in the pregnancies in question. This latter feature considerably simplifies the type of PCR assay that can be used.

In these preliminary studies [4], we examined 22 plasma samples obtained from rhesus D pregnant women. The plasma DNA was extracted from 400–800 μl plasma using Qiagen column technology, as we have found this to give very good yields and a optimal template for the subsequent PCR reactions. We have observed similarly good results using Roche column technology. All PCR reactions were performed in a branched nested manner, in that all loci were initially amplified in a multiplex PCR, after which each specific locus was then further examined in a separate nested PCR. To ensure that DNA was indeed present in the sample being examined, we used the ubiquitous β-globin gene. In this manner, we could ensure that erroneous results were not obtained due to technical drop outs. This study [4] showed that such an approach can indeed be used for the effective detection of multiple fetal loci on cell free DNA in maternal DNA, in that we were able to correctly determine both fetal loci with almost 100% accuracy. In only one instance did we record an incorrect result, which was probably attributable to contamination; a potentially great problem when using such highly sensitive nested PCR protocols.

For this reason and since this assay was too cumbersome to be used for routine diagnostic applications, we elected to develop a multiplex TaqMan assay for these two fetal loci. The TaqMan real-time PCR system offers distinct advantages over conventional PCR systems. Firstly, as the product generated by the PCR is recognized by a highly specific oligonucleotide probe, this raises the specificity of the assay considerably, since PCR artefactual by-products are no longer of any concern. Secondly, the online real-time monitoring of the PCR via the elimination of the specific probe means that no more tubes need to be opened in order to assess the PCR, for instance by gel electrophoresis. Hence, potential contamination by any PCR products is greatly reduced by the use of this novel closed system. Thirdly, since the PCR assay is actually monitored cycle for cycle in a real-time manner, it is also possible to quantitate the amount of the input template. This becomes important when considering other clinical implications as outlined below. A further important factor is that this assay is much more amenable to automation than conventional PCR and that the format used permits the analysis of up 96 different samples per analytic run. Consequently, the use of an appropriate multiplex assay permits one to examine both fetal loci on line in the same tube in the same reaction.

In our hands this system has proven to be very effective and we have obtained results very analogous to those previously achieved with the nested conventional PCR assay, in that of the 34 samples examined 12/13 rhesus D males were correctly identified, all 5 rhesus D males were correctly identified and the rhesus D genotype of all 16 female fetuses was correctly identified.

Despite these promising results we would like to caution against the premature application of this technology in the clinical setting as in an exhaustive analysis of over 200 samples we obtained false-negative rates of close to 6% [5]. These failures cannot be attributed to inadequate amounts of free DNA being present in the sample, as simultaneous measurement of the total amount of free DNA, of which the vast majority is derived from the mother, indicates that there were copious amounts of DNA present in the samples in question. Furthermore as they were obtained late in pregnancy, at which time there is a general significant increase in the concentration of free fetal DNA, it cannot be argued that there may have been insufficient material present due to the samples being taken at too early a gestational age.

Rather, it seems that adequate strategies should be adopted to counter such potential false-negative results. One obvious measure is to take the approach we have used and ascertain that there is indeed DNA present in the sample being analyzed by using a ubiquitous genetic locus, such as β-globin or GAPDH. The other more specific measure would be to include an analysis of paternally inherited highly polymorphic markers, such as short-tandem repeats and thereby ensure that there is indeed some fetal DNA present in the sample being examined. While not as sensitive as a TaqMan assay, the use of fluorescent quantitative PCR should be able to detect fetal DNA quantities in the range normally present (approximately 1% of the total DNA present).

Free Fetal DNA Levels in Pregnancies with Aneuploid Fetuses

Earlier studies examining enriched fetal cells found that greater numbers of fetal cells may be present in the periphery of pregnant women with aneuploid fetuses, in that considerably more cells with triploid chromosome-specific FISH signals are observed in such pregnancies than in those with karyotypically normal fetuses [6, 7].

This aspect was confirmed in a more defined manner in a study carried out by Bianchi et al. [8] who, by using a quantitative PCR based on the incorporation of radioactive nucleotides into the PCR product, showed that approximately 5-fold more fetal cells were present in trisomy 21 fetuses but not in those with trisomy 18. Some elevation was also noted in fetuses with trisomy 13.

As elevations in both fetal cell traffic and free fetal DNA had been noted in pregnancies with preeclampsia [9], we were curious to see if free fetal DNA levels are also elevated in a similar manner to fetal cell levels in pregnancies with aneuploid fetuses. As a large number of plasma specimens from patients with a high risk for carrying an aneuploid fetus had been collected in the NIH NIFTY study, our group in collaboration with Laird Jackson [10] in Philadelphia examined free fetal DNA levels in a series of such affected pregnancies. These included 15 pregnancies with male trisomy 21 fetuses and 6 with trisomy 18. Our analysis demonstrated that the mean free fetal DNA levels were indeed significantly elevated by approximately 2-fold in fetuses with trisomy 21 when compared to the matched healthy control cohort [10]. No such elevation was observed in those pregnancies with trisomy 18 fetuses. These data are in good accordance with a similar report made by Lo et al. [11], in which they also observed a similar 2-fold elevation in free fetal DNA levels in pregnancies with trisomy 21 fetuses. Although we did observe some increase in free fetal DNA levels for trisomy 13, our sample size was too small to draw any concrete conclusions. In this regard our data are remarkably similar to those made independently for fetal cell levels by Bianchi et al. [8], in that elevations for both parameters are seen in trisomy 21 fetuses but not in trisomy 18 fetuses. The difference in the levels of fetal circulatory DNA for fetuses with trisomy 21 compared to trisomy 18 could in part be explained by the placental abnormalities associated with these chromosomal aberrations, in that the placental and fetal mass of 21-trisomic infants is similar to that of normal fetuses whereas those of infants with trisomy 18 are reduced and growth retarded [12].

Although our study does show a significant difference between normal pregnancies and those with an aneuploid fetus, particularly those with trisomy 21, a larger scale study is necessary to determine the true extent of this finding. A clear drawback of the current technology is that it is focused on detecting male DNA sequences, due to the ease of this assay. This deficit could, however, be readdressed by examining paternally inherited highly polymorphic sequences, as recently described [13]. By the application of such developments, it will be interesting to determine whether quantification of fetal DNA levels can be included in an integrated screening approach to improve upon the current low levels of specificity and sensitivity for detecting affected fetuses.

Free DNA Levels in Preeclampsia

A novel finding of our laboratory was that fetal cell traffic into the maternal periphery was significantly enhanced in pregnancies affected by preeclampsia [14]. By the use of a real-time TaqMan assay, Lo et al. [15] were able to show that free fetal DNA was also similarly significantly elevated in preeclamptic pregnancies.

In order to confirm this latter observation, in an independent study we performed a large-scale blinded investigation in collaboration with Baha Sibai and Jeff Livingston (Memphis, Tenn., USA) and Olavi Ylikorkala and Hannele Laivouri (Helsinki, Finland) [16]. This examination showed that free fetal DNA levels were indeed significantly elevated almost 10-fold under conditions of preeclampsia. However, as we were also monitoring the total levels of free DNA by the use of a TaqMan assay for the ubiquitous GAPDH gene, we noticed that these levels were also similarly significantly elevated to the same extent. As the vast majority of the free DNA in the maternal circulation is derived from the mother (>95%), this new result indicated that free maternal DNA is also elevated in preeclampsia. Since it is likely that this increase in the release of free DNA is due to some form of cellular damage or death, this result implies that in preeclampsia there is an insult to both the fetal and maternal compartments. A further noteworthy finding was that the increments in free DNA levels corresponded to the degree of severity and were correlated to each other. This latter feature was only evident in pregnancies with preeclampsia but not in normal healthy pregnancies.

These data fit very well to a model whereby damage in a fetal compartment leads to damage in the mother, and thereby unifies theories put forward by Redman and Roberts [17, 18] whereby the release of microparticles by the destruction of the placenta lead to widespread damage of the maternal endovascular system.

As we had recently performed a large-scale prospective study to study whether fetal cell traffic into the maternal circulation was already altered early during pregnancy, we examined the plasma DNA free levels in these samples. Our analysis of these samples showed that free fetal DNA levels were already elevated approximately 3-fold at 20 weeks of pregnancy in those pregnancies which subsequently developed preeclampsia when compared to those with normal obstetrical outcome. No distinct difference was noted in the levels of free maternal DNA between the study group and the control cohort. Akin to what we had observed for fetal cell traffic, no significant elevation was observed in those women who gave birth to growth-retarded fetuses.

This result implies that while the lesion leading to the increased release of free fetal DNA occurs early in the development of preeclampsia, the lesion associated with the increased release of free maternal DNA only occurs much later, perhaps only once the symptoms have become fulminant.

Free Fetal DNA Levels and Fetal Cell Levels Do Not Correlate

Since we have observed elevations in fetal cell traffic and the release of free fetal DNA into the maternal circulation, we asked whether these two parameters were correlated or whether they occurred independent of each other. For this study we measured the amount of free fetal DNA using our TaqMan assay, whereas the number of fetal cells was determined by enrichment and subsequent FISH or single cell PCR as the TaqMan assay is not sensitive enough to detect rare fetal cells directly in the maternal circulation.

Our data, obtained both in normal pregnancies and those with preeclampsia, showed that there was no significant correlation between the levels of free fetal DNA and fetal cells in maternal blood. This result suggests that these two parameters are regulated independent of each other. A further possibility is that, since fetal erythroblasts persist for much longer in the maternal circulation (almost 90 days) whereas free fetal DNA is very short lived (approximately 15 min), these two events may appear to happen asynchronously.

Fluctuation of Free DNA Levels

As our data using real-time PCR have suggested that elevated levels of free fetal DNA occur in fetuses with trisomy 21, they may provide the basis for a new screening tool. A premise for such assays would be that the levels of the analyte being measured are stable over the time of the analyses, and are not subject to any large fluctuations. Although it has been determined that free fetal DNA levels have a very short half life in the maternal circulation, no study to date has addressed whether their levels are stable or fluctuate. In our study [19], we first examined total free DNA in normal healthy male and female (nonpregnant) individuals over a period of up to 6 days. This analysis showed that up to 13.5-fold variations could be determined in samples taken twice daily over this period. Astounded by this feature we next examined pregnant women with male fetuses. Here samples were obtained once daily over a period of 3 days. Again here we observed large variations in the amount of maternal free DNA, with fluctuations of up to 21.5-fold being recorded. Although the amount of free fetal DNA did not vary to the same extent, we did record variations of up to 2.2-fold. As the levels of these variations are of the same order as the elevations seen in aneuploid fetuses, our data do suggest that considerable caution should be taken when considering this methodology as a potential screening tool.

Conclusions

The discovery of free fetal DNA in the maternal circulation has opened a new avenue for the facile detection of certain fetal genetic loci. A limit of this technology is that since maternal free DNA is also present in the maternal plasma, only those paternally inherited genes absent from the maternal genome can be detected with relative ease. Nevertheless, this technology could permit the rapid and simple assessment of the fetal rhesus D status and sex in pregnancies at risk for a X-linked disorder. The quantitation of this material using sensitive real-time PCR assays has yielded new insights into a variety of pregnancy-associated disorders, in particular preeclampsia, where our data indicate that both the maternal and fetal compartments are affected in a manner proportional to disease severity.

The usefulness of this technology, for instance as a potential screening tool for the detection of fetal aneuploidies, may be hindered by the considerable fluctuations we have found to occur under normal conditions.

Acknowledgements

This work was supported in part by the Swiss National Science Foundation, grant number 3200-047112.96, and the NIH (USA), contract number N01-HD-4-3202.

References

1 Lo YMD, Corbetta N, Chamberlain PF, Sargent JL: Presence of fetal DNA in maternal plasma and serum. Lancet 1997;350:485–487.
2 Lo YM: Fetal DNA in maternal plasma. Ann NY Acad Sci 2000;906:141–147.
3 Troeger C, Zhong XY, Burgemeister R, Tercanli S, Minderer S, Holzgreve W, Hahn S: Approximately half the erythroblasts in maternal blood are of fetal origin. Mol Hum Reprod 1999;5:1162–1165.
4 Zhong XY, Holzgreve W, Hahn S: Detection of fetal rhesus D and sex from fetal DNA in maternal plasma by multiplex PCR. Br J Obstet Gynaecol 2000;107:766–769.
5 Zhong XY, Hahn S, Holzgreve W: Prenatal analysis of fetal genetic traits. Lancet 2001;357:310–311.
6 Price JO, Elias S, Wachtel SS, Klinger K, Dockter M, Tharapel A, Shulman LP, Phillips OP, Meyers CM, Shook D, Simpson JL: Prenatal diagnosis with fetal cells isolated from maternal blood by multiparameter flow cytometry. Am J Obstet Gynecol 1991;165:1731–1737.
7 Gänshirt-Ahlert D, Börejesson-Stoll R, Burschyk M, Dohr A, Gerritsen HSP, Helmer E, Miny P, Velasco M, Walde C, Patterson D, Teng N, Bhat NM, Bieber MM, Holzgreve W: Detection of fetal trisomies 21 and 18 from maternal blood using triple gradient and magnetic cell sorting. Am J Reprod Immunol 1993;30:194–201.
8 Bianchi DW, Williams JM, Sullivan LM, Hanson FW, Klinger KW, Shuber AP: PCR quantitation of fetal cells in maternal blood in normal and aneuploid pregnancies. Am J Hum Genet 1997;61: 822–829.
9 Holzgreve W, Hahn S: Novel molecular biological approaches for the diagnosis of preeclampsia. Clin Chem 1999;45:451–452.
10 Zhong XY, Bürk MR, Troeger C, Jackson LR, Holzgreve W, Hahn S: Fetal DNA in aneuploid fetuses is elevated in pregnancies with aneuploid fetuses. Prenat Diagn 2000;20:795–798.
11 Lo YM, Lau TK, Zhang J, Leung TN, Chang AM, Hjelm NM, Elmes RS, Bianchi DW: Increased fetal DNA concentrations in the plasma of pregnant women carrying fetuses with trisomy 21. Clin Chem 1999;45:1747–1751.
12 Creasy R, Resnik R: Intrauterine growth restriction; in Creasy R, Resnik R (eds): Maternal Fetal Medicine, ed 4. Philadelphia, Saunders, 1999, pp 569–584.
13 Tang NL, Leung TN, Zhang J, Lau TK, Lo YM: Detection of fetal-derived paternally inherited X-chromosome polymorphisms in maternal plasma. Clin Chem 1999;45:2033–2035.
14 Holzgreve W, Ghezzi F, Di Naro E, Maymon E, Gänshirt D, Hahn S: Disturbed fetomaternal cell traffic in preeclampsia. Obstet Gynecol 1998;91:669–672.
15 Lo YM, Leung TN, Tein MS, Sargent IL, Zhang J, Lau TK, Haines CJ, Redman CW: Quantitative abnormalities of fetal DNA in maternal serum in preeclampsia. Clin Chem 1999;45:184–188.
16 Zhong XY, Laivuori H, Livingston JC, Ylikorkala O, Sibai BM, Holzgreve W, Hahn S: Elevation of both maternal and fetal extracellular circulating deoxyribonucleic acid concentrations in the plasma of pregnant women with preeclampsia. Am J Obstet Gynecol 2001;184:414–419.
17 Redman CW, Sacks GP, Sargent IL: Preeclampsia: an excessive maternal response to pregnancy. Am J Obstet Gynecol 1999;180:499–506.
18 Roberts JM, Cooper DW: Pathogenesis and genetics of preeclampsia. Lancet 2001;357:53–56.
19 Zhong XY, Burk MR, Troeger C, Kang A, Holzgreve W, Hahn S: Fluctuation of maternal and fetal free extracellular circulatory DNA in maternal plasma. Obstet Gynecol 2000;96:991–996.

Dr. Sinuhe Hahn, Department of Obstetrics and Gynecology, University of Basel
Schanzenstrasse 46, CH–4031 Basel (Switzerland)
Tel. +41 325 92 24, Fax +41 325 93 99, E-Mail shahn@uhbs.ch

Hahn S, Holzgreve W (eds): Fetal Cells and Fetal DNA in Maternal Blood. New Developments for a New Millennium. 11th Fetal Cell Workshop, Basel, 2000. Basel, Karger, 2001, pp 104–110

Quantitative Analysis of Fetal DNA in Maternal Plasma in Preeclamptic Pregnancies

L. Cremonesi[a], *A. Vassallo*[a], *M. Ferrari*[a], *F. Lagona*[b], *L. Valsecchi*[b], *A. Ferrari*[b], *M. Smid*[b]

[a] Unit of Genetics and Molecular Diagnostics and
[b] Department of Obstetrics and Gynecology, University Hospital San Raffaele, Milan, Italy

Introduction

In the past, fetal DNA has been extracted from nucleated blood cells in maternal peripheral blood [1, 2]. More recently, the presence of free fetal DNA circulating in maternal plasma has also been demonstrated [3]. In order to differentiate the fetal genome from the maternal one, PCR amplification of Y-chromosome sequences from a male fetus has been employed as a model system, since this sequence is lacking in the mother. Different approaches have been developed: initially, a nested PCR assay with a high number of cycles has been carried out on DNA extracted from nucleated cells of maternal blood [1, 2], while a single PCR assay to amplify fetal DNA in maternal plasma has been described [3]. In order to evaluate the diagnostic accuracy of these protocols and to identify the optimal source for fetal genetic material, we compared single PCR on plasma DNA versus nested PCR on DNA extracted from either plasma or nucleated blood cells [4, 5]. This study revealed that nested PCR on fetal DNA from maternal plasma produced the highest number of correct fetal sex identifications and displayed the best combination of high sensitivity (100%), specificity (91%), accuracy (96%), positive (94%) and negative predictive values (100%) in fetal gender determination compared to the other approaches. This study also revealed that the plasma approach appeared more promising than the use of nucleated blood cell DNA, independent of the amplification protocol employed for a future application to noninvasive prenatal diagnosis.

The presence of fetal DNA in maternal plasma has only recently been explored and data about factors affecting its release and clearance into and from maternal circulation are still scarce.

Existing data on the variation of fetal plasma DNA concentrations are scanty and show a progressive increase throughout pregnancy [6]. This is in contrast to our previous findings on fetal DNA retrieval in maternal whole blood, which drops from the first to the second trimester [7]. Further data are needed on the range and distribution of fetal DNA in maternal plasma at different gestational ages in physiological pregnancies.

Furthermore, the placental pathological conditions described in pregnancies complicated by hypertension and preeclampsia seem to be associated with an enhanced cell passage into maternal circulation and higher fetal DNA concentrations in maternal plasma [8, 9].

We wanted to confirm these data and investigate whether abnormal patterns of fetal DNA in maternal plasma can be detected even before clinical signs. This might result in a powerful test for the risk of developing this pathology, allowing the adoption of preventive strategies. For this purpose, we performed real-time quantitative PCR on the SRY sequence in physiological and pathological pregnancies in women carrying male fetuses [6]. We also wanted to evaluate whether the fetal DNA concentration correlates with other biochemical and biophysical markers of preeclampsia (plasmatic fibronectin levels [10] and Doppler velocimetry of the uterine arteries [11]), in order to verify whether these parameters can be used in combination as early predictors of the development of pathological placental conditions.

Subjects and Methods

To determine the range and distribution of the fetal DNA concentration in maternal plasma, we collected blood samples from an unselected population of pregnant women attending the ambulatory care facilities and the prenatal diagnosis unit of our hospital. Fetal gender was unknown at the time of blood sampling and was confirmed at delivery or by karyotyping. In order to evaluate whether in placental pathologies there is an increase in fetal DNA release into maternal circulation, we collected blood samples from either pregnant women at a high risk of developing preeclampsia, based on clinical and anamnestic findings, or from pregnant women who had already developed symptoms at the time of sampling.

Sample Preparation

Ten milliliters of maternal peripheral blood were collected in EDTA vacutainer tubes. Blood was immediately iced and processed within a few hours after venesection. After centrifugation at 3,000 *g* at 4 °C, plasma was carefully removed and recentrifuged at 3,000 *g* and the supernatant was collected into fresh tubes. Samples were stored at –20 °C until further processing.

DNA Extraction from Plasma

DNA was extracted from plasma samples with the QIAamp Blood Kit (Qiagen) according to the protocol recommended by the manufacturer. We used 400 μl of the plasma sample per column for DNA extraction.

Fibronectin Determination

Fibronectin levels in maternal plasma were determined through immunonephelometry with a Nephelometer Analyzer II, DADE Behring. Normal values were considered those within the range of 0.25–0.4 g/l.

Doppler Velocimetry of the Uterine Arteries

Doppler velocimetry of the uterine arteries was carried out with a Hitachi Eidos 3.75 MHz probe according to Bewley et al. [11].

Real-Time TaqMan Quantitative PCR

Real-time quantitative PCR analysis was performed with an ABI Prism 7700 Sequence Detector System (PE Applied Biosystems) [12]. The amplification and product-reporting system used is based on the 5′-nuclease assay [13]. In this system, in addition to the two amplification primers, as in conventional PCR, a dual-labeled fluorogenic hybridization probe is also included. One fluorescent dye (6 carboxyfluorescein) serves as a reporter and its emission spectrum is quenched by a second fluorescent dye (6-carboxytetramethylrhodamine). During the extension phase of PCR, the 5′- to 3′-exonuclease activity to the Taq DNA polymerase cleaves the reporter from the probe, thus releasing it from the quencher, resulting in an increase in fluorescent emission. The release of a fluorogenic tag from the 5′ end of the probe is proportional to the target sequence concentration (copy number).

In order to amplify fetal Y-specific sequences, we used primers SRY-109F, SRY-245R and probe SRY-142T′ as described [6]. As a control for the presence of DNA, we amplified the β-globin gene by using primers β-globin-354F, β-globin-455R, and the probe β-globin-402T [6]. Reactions were carried out in a volume of 50 μl by using components (except probes and primers) supplied in a TaqMan 2 × PCR Master Mix (PE Biosystems). We used 10 μl of extracted plasma DNA for amplification. Each sample was analyzed in quadruplicate. A calibration curve was run in parallel and in duplicate with each analysis. Thermal cycling consisted of a 2-min incubation at 50 °C to allow the anti-contamination uracil N-glycosylase system to act, followed by a first denaturation step of 10 min at 95 °C and then 40 cycles of 95 °C for 15 min and 60 °C for 1 min. The same thermal cycling conditions were used for both SRY and the β-globin TaqMan systems [6].

Results

In a sample of 44 control physiologically pregnant women carrying male fetuses, fetal DNA in maternal plasma varied from 22.67 genome equivalents (g.e.)/ml (mean value) during the first trimester (range 0–89.9 g.e./ml) to 20.94 (mean value) during the second trimester (range 0–82.3 g.e./ml), and to 42.26 (mean value) during the third trimester (range 0–164.1 g.e./ml).

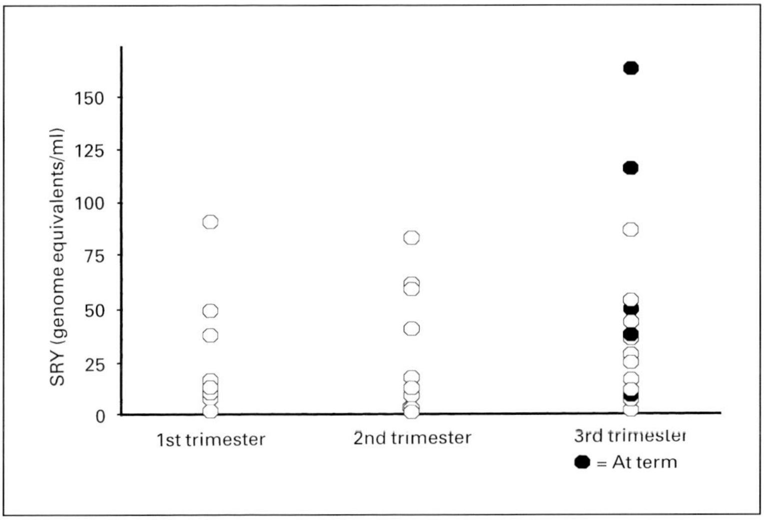

Fig. 1. Fetal DNA concentration in maternal plasma in physiological pregnancies sampled in the three gestational trimesters.

Figure 1 shows the fetal DNA concentration in maternal plasma determined in physiological pregnancies during the three gestational trimesters, and at term.

In table 1, the fetal DNA concentration, evaluation of the Doppler velocimetry of the uterine arteries and fibronectin levels in maternal plasma are shown for 6 pregnant women with hypertension and either a physiological or pathological outcome, and in 5 women with overt preeclampsia.

Figure 2 illustrates fetal DNA quantification in the maternal plasma of 5 pregnant women with overt manifestations of preeclampsia and 4 pregnant women analyzed before manifestation of preeclampsia, compared to the group of control pregnant women with physiological gestation sampled during the third trimester.

Discussion

We evaluated a cohort of physiological pregnancies examined during the three trimesters of gestation. The fetal DNA concentration in maternal plasma was comparable during the three trimesters except for the samples obtained from five gestations at term, who displayed the highest fetal DNA concentration (fig. 1).

Interestingly, fetal DNA in 5 preeclamptic pregnancies, sampled during the third trimester, had a higher concentration compared with the values obtained for

Table 1. Markers of preeclampsia

Sample No.	Gestational week	Diagnosis	Doppler velocimetry	Fibronectin g/l	Fetal DNA g.e./ml
1	35+5	Hypertension	Normal	0.42	21.8
2	28	Hypertension	Normal	0.29	3.8
3	25+2	Hypertension	Abnormal	0.46	26.5
3	30+2	Hypertension (preeclampsia at 34 weeks)	Abnormal	0.29	118.6
4	24+4	Hypertension (preeclampsia at 37 weeks)	Abnormal	0.36	35.32
5	13	Hypertension	Normal	0.46	5.8
5	24	Hypertension (preeclampsia at 28 weeks)	Abnormal	0.60	13.5
6	24	Hypertension	Abnormal	0.23	82.3
6	29	Hypertension (preeclampsia at 37 weeks)	Abnormal	0.32	25.42
7	34	Overt preeclampsia	Normal	0.93	568
8	35	Overt preeclampsia, IUGR	Abnormal	0.51	473.5
9	34	Overt preeclampsia	Abnormal	0.91	146.7
10	33	Overt preeclampsia, IUGR	Abnormal	0.58	512.3
11	34	Overt preeclampsia, IUGR	Abnormal	0.88	119.5

Quantification of the fetal DNA concentration in maternal plasma, Doppler velocimetry of the uterine arteries and fibronectin levels in maternal plasma of 2 women with mild hypertension who did not develop preeclampsia (samples 1 and 2), in 4 women who developed preeclampsia (samples 3–6) and in 5 women with overt preeclampsia (samples 7–11). IUGR = Intrauterine growth retardation; n.d. = not determined.

control pregnancies during the same trimester (fig. 2; table 1). In contrast, fetal DNA was elevated only in 1 of 4 pregnant women who developed preeclampsia some weeks later (fig. 2; table 1).

Fetal DNA levels within the physiological range were also found in 2 pregnant women at high risk of preeclampsia, who later developed only a mild, non-proteinuric gestational hypertension with uneventful pregnancy (table 1, samples 1 and 2).

We also investigated whether a relationship exists between the fetal DNA concentration and some biochemical and biophysical findings, in an effort to correlate these parameters with the development of preeclampsia. Table 1 shows that these values were normal in the 2 women displaying only mild hypertension not complicated by preeclampsia (table 1, samples 1 and 2). Similarly, among the 4

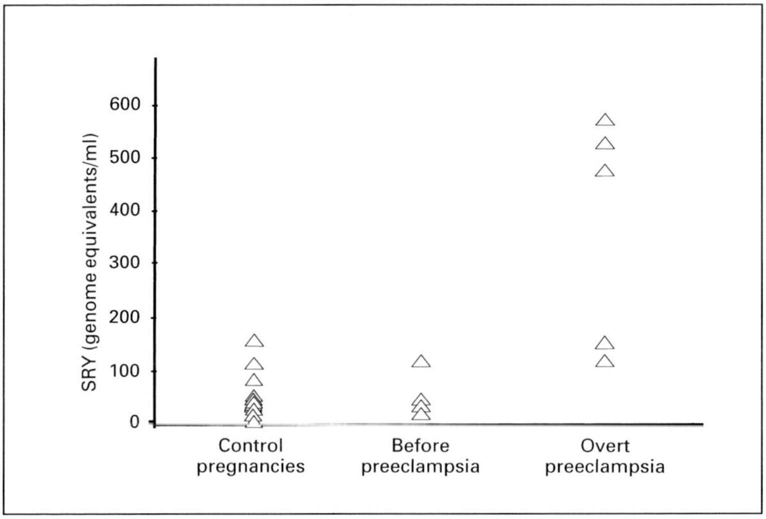

Fig. 2. Fetal DNA concentrations in maternal plasma in a group of control physiological pregnancies sampled in the third trimester of gestation, compared with a group of pregnant women with hypertension who developed preeclampsia some weeks later and a group of pregnant women with overt preeclampsia.

women displaying hypertension who developed preeclampsia some weeks later (table 1, samples 3–6), the fetal DNA concentration was elevated in only 1 case about 4 weeks before the pathological outcome. In the same women, fibronectin levels were abnormal in only in 1 case, while the Doppler velocimetry of the uterine arteries was pathological in all pregnancies. Conversely, in pregnant women displaying overt preeclampsia (table 1, samples 7–11), the fetal DNA concentration was much higher than in physiological conditions, as well as fibronectin levels. In the same cases, Doppler velocimetry of the uterine arteries was abnormal in 4 of 5 cases.

Our findings show that isolated nonproteinuric hypertension does not seem to correlate with an increase in fetal DNA in maternal plasma. In contrast, an overt clinical manifestation of preeclampsia is associated with a marked increase in fetal DNA. This rise may precede clinical signs by only a few weeks. Further extensive longitudinal studies are needed to determine the timing of a detectable rise in fetal DNA levels in maternal plasma before overt clinical manifestations of preeclampsia in order to evaluate whether fetal DNA quantification may be used as an early predictor of the development of this and other placental pathological conditions.

References

1 Lo YMD, Patel P, Sampietro M, Gillmer MDG, Fleming KA, Wainscoat JS: Detection of single-copy fetal DNA sequence from maternal blood. Lancet 1990;335:1463–1464.

2 Lo YMD, Patel P, Baigent CN, Gillmer MDG, Chamberlain P, Travi M, Sampietro M, Wainscoat JS, Fleming KA: Prenatal sex determination from maternal peripheral blood using the polymerase chain reaction. Hum Genet 1993;90:483–488.

3 Lo YMD, Corbetta N, Chamberlain PF, Sargent RV, Redman CWG, Wainscoat JS: Presence of fetal DNA in maternal plasma and serum. Lancet 1997;350:485–487.

4 Smid M, Lagona F, De Benassuti L, Ferrari A, Ferrari M, Cremonesi L: Evaluation of different approaches for fetal DNA analysis from maternal plasma and nucleated blood cells. Clin Chem 1999;45:1570–1572.

5 Lagona F, Smid M, de Benassuti L, Ferrari A, Ferrari M, Cremonesi L: Fetal DNA analysis from maternal plasma and nucleated blood cells. Ann NY Acad Sci 2000;906:156–160.

6 Lo YMD, Tein MSC, Lau TK, Haines CJ, Leung TN, Poon PMK, Wainscoat JS, Johnson PJ, Chang AMZ, Hjelm NM: Quantitative analysis of fetal DNA in maternal plasma and serum: Implications for noninvasive prenatal diagnosis. Am J Hum Genet 1998;62:768–775.

7 Smid M, Lagona F, Papasergio N, Ferrari A, Ferrari M, Cremonesi L: Influence of gestational age on fetal deoxyribonucleic acid retrieval in maternal peripheral blood. Am J Obstet Gynecol 1997;177:517–522.

8 Lo YMD, Leung TN, Tein MSC, Sargent IL, Zhang J, Lau TK, Haines CJ, Redman CWG: Quantitative abnormalities of fetal DNA in maternal serum in preeclampsia. Clin Chem 1999;45:184–188.

9 Holzgreve W, Ghezzi F, Di Naro E, Gaenshirt D, Maymon E, Hahn S: Disturbed feto-maternal traffic in preeclampsia. Obstet Gynecol 1998;91:669–672.

10 Lazarchick J, Stubbs T, Romein L, van Dorsten JP, Loadmolt CB: Predictive value of fibronectin levels in normotensive gravid women destined to become pre-eclamptic. Am J Obstet Gynecol 1986;154:1050–1056.

11 Bewley S, Cooper D, Campbell S: Doppler investigation of uteroplacental blood flow resistance in the second trimester: A screening study for pre-eclampsia and intrauterine growth retardation. Br J Obstet Gynaecol 1991:88:871–879.

12 Heid CA, Stevens J, Livak KJ, Williams PM: Real time quantitative PCR. Genome Res 1996;6:986–994.

13 Holland P, Abramson RD, Watson R, Gelfand DH: Detection of specific polymerase chain reaction product by utilizing the 5′-3′ exonuclease activity of *Thermus aquaticus* DNA polymerase. Proc Natl Acad Sci USA 1991;88:7276–7280.

Dr. Laura Cremonesi, Unit of Genetics and Molecular Biology, University Hospital San Raffaele
Via Olgettina 58, I–20132 Milan (Italy)
Tel. +39 02 26434779, Fax +39 02 26434767, E-Mail cremonesi.laura@hsr.it

Hahn S, Holzgreve W (eds): Fetal Cells and Fetal DNA in Maternal Blood. New Developments for a New Millennium. 11th Fetal Cell Workshop, Basel, 2000. Basel, Karger, 2001, pp 111–119

Circulating Trophoblast Cells in Maternal Peripheral Blood

I.J. van Wijk[a], *M.L. Tjoa*[a], *S. Griffioen*[a], *M.A.M. Mulders*[a], *J.M.G. van Vugt*[b], *C.B.M. Oudejans*[a]

[a] Molecular Biology Laboratory, Department of Clinical Chemistry, and
[b] Prenatal Diagnostics Center, Department of Obstetrics and Gynecology, University Hospital Vrije Universiteit, Amsterdam, The Netherlands

Trophoblast Cells

Trophoblast cells are invasive extra-embryonic cells, essential for implantation, for the establishment of functional connections with the vascular system of the mother and for subsequent growth and development of the embryo and placenta. During early pregnancy trophoblast cells invade maternal stromal tissue (interstitial invasion) and uterine spiral arteries and a subset of veins [1, 2] (endovascular invasion). During this process, endovascular trophoblast cells actively and selectively migrate into the maternal vessels in a retrograde fashion thereby replacing endothelial cells [1, 3]. Endovascular trophoblast invasion results in a modification of the maternal vessels generating large-bore/low-resistance vessels that guarantee a constant and optimal blood flow from the mother to the growing fetus. This physiological change of spiral arteries only occurs in the presence of trophoblasts [1, 4]. Trophoblast dysfunction, presented by reduced trophoblast invasion and resulting in the absence of spiral artery modification, may lead to miscarriage, preeclampsia and/or intrauterine growth retardation of the fetus [5–7]. An abnormality in the immunosuppressive process, as the maternal immune system encounters paternal antigens, might cause the deficient invasion. Indeed Goldman-Wohl et al. [8] showed downregulation of HLA-G expression in extravillous trophoblast cells in preeclamptic patients. However, recently a null mutation in HLA-G was shown not to be associated with an increased incidence of preeclampsia or intrauterine growth retardation in pregnancy [9]. Another explanation for the lack of trophoblast invasion seen in these pregnancy disorders might be incomplete differentiation of the extravillous trophoblast cells, either intrinsically programmed by a genetic defect or as a consequence of a defective interaction with maternal cells. In normal pregnancy the expression profile of endovascular trophoblast cells shows a switch during invasion from an epithelial to a mesenchymal phenotype [3] that is essential for and accompanies the process of invasion. However, in preeclampsia this switch does not occur and trophoblast cells continue to express cell adhesion molecules characteristic of the epithelial phenotype [10, 11]. Thus in preeclamptic patients the phenotype of extravillous trophoblast cells differs drastically from that of normal pregnancies, although it remains to be proven whether this failure is a cause or consequence.

In healthy pregnancies, endovascular trophoblast cells are dragged by the maternal blood flow and circulate in maternal peripheral blood in early gestation. These fetal cells are good candidates for target cells for noninvasive prenatal diagnosis and, as described above, might also be indicative of trophoblast function, and thus placental (mal)function. Phenotypic analysis of these cells following enrichment from the maternal blood could permit prediction of pregnancies at risk for preeclampsia or other abnormalities associated with trophoblast dysfunction as early as in the first trimester of pregnancy.

The Trophoblast Cell as a Target Cell for Noninvasive Prenatal Diagnosis

As trophoblast cells appear very early in the maternal peripheral circulation, are not present in normal adult blood and probably do not persist after delivery, these fetal circulating cells are, like fetal nucleated erythrocytes [12], good candidates for noninvasive prenatal diagnosis. Thomas et al. [13], detected fetal DNA in maternal peripheral blood starting at 5 weeks of gestation. At this stage, fetal nucleated red blood cells (NRBCs) have not yet been formed. Consequently, fetal DNA in maternal blood at this early stage is most probably derived from circulating trophoblast cells. The trophoblast cells we have chosen to isolate from maternal blood are the uninuclear cytotrophoblast cells, derived from the population of

Fig. 1. Flow diagram of the isolation and analysis of trophoblast cells and plasma from maternal peripheral blood during early pregnancy. Twenty-five milliliters of peripheral heparin blood is loaded on a 5-layer discontinuous Percoll gradient. After centrifugation plasma is used for X/Y PCR and FISH analysis. Cells from the gradient are used either (i) for a second enrichment step by miniMACS, followed by X/Y PCR and by HASH2 RNA in situ hybridization, or (ii) directly after centrifugation for combined immunocytochemistry and FISH (noninvasive prenatal diagnosis) or for double-label immuno-identification and immuno-phenotyping (early identification of pregnancies at risk of preeclampsia and/or intrauterine growth retardation).

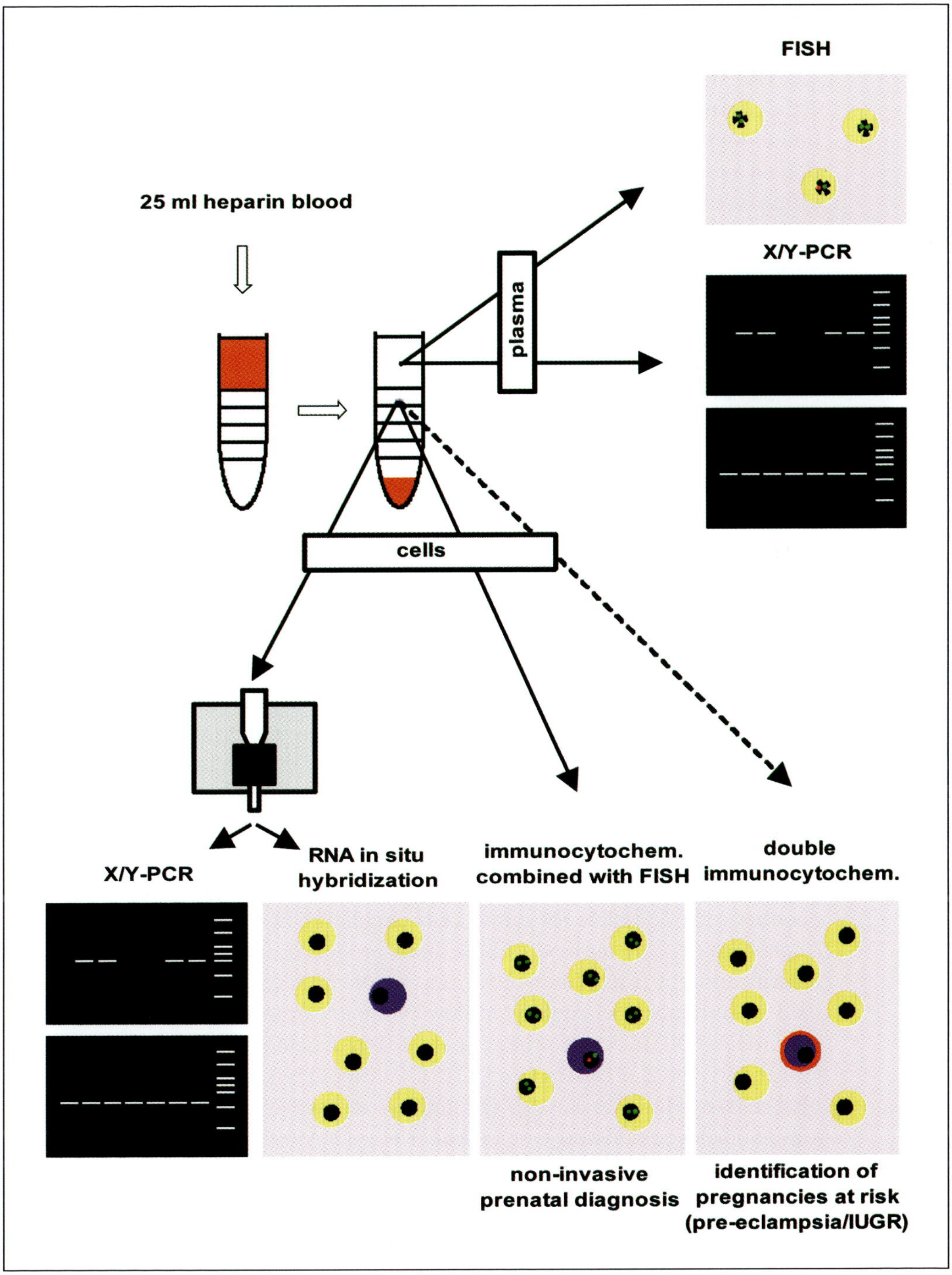
FISH
25 ml heparin blood
X/Y-PCR
plasma
cells
X/Y-PCR
RNA in situ hybridization
immunocytochem. combined with FISH
double immunocytochem.
non-invasive prenatal diagnosis
identification of pregnancies at risk (pre-eclampsia/IUGR)

extravillous trophoblast cells (i.e. intermediate trophoblast cells) which invade maternal spiral arteries [1] and a subset of veins as suggested recently [2].

Our first approach was a two-step procedure for enrichment of trophoblast cells, consisting of a density gradient separation followed by miniMACS depletion of maternal white blood cells [14] (fig. 1). Cells are separated by centrifugation on a discontinuous 5-layer Percoll gradient. Subsequently, additional purification is performed through depletion by magnetic activated cell sorting of CD50-positive maternal cells. This antigen is present on more than 95% of all peripheral leukocytes and absent on trophoblast cells. To demonstrate that in the enriched cell fractions fetal DNA was indeed present, we have developed a highly sensitive and specific nested PCR for amplification of X- and Y-chromosome specific sequences [14]. For detection of Y-chromosomal DNA, primers specific for the single-copy gene SRY were used. Amplification of an X-chromosome specific sequence was performed as an internal control for the presence and for correct amplification of DNA. Fetal sex prediction using this X/Y PCR on enriched samples from peripheral blood of 36 pregnant women resulted in 92% correct predictions (33/36) [14]. These results imply that the sequential application of a density gradient centrifugation and miniMACS protocol enables selection for fetal trophoblast cells, suitable for further genetic analysis, for instance by fluorescent DNA in situ hybridization (FISH).

Using this two-step enrichment procedure, still numerous maternal cells are present among the selected fetal cells. As the number of trophoblast cells is very low, comparable to other fetal cells (NRBCs) circulating in maternal blood [12, 15], reliable identification of these cells is extremely important. Frequently cytokeratin is used as a marker for trophoblast cells as for other cells of epithelial origin. However, our own experience as well as that of others focusing on identification of circulating tumor cells, show that low-level expression of epithelium markers by hematopoietic cells is present and detectable [16] and as such leads to false positivity. Therefore other markers should be used for identification of circulating trophoblast cells as well as for tumor cells circulating in peripheral blood. Various trophoblast-specific antibodies have been produced and used for isolation and identification of trophoblast cells from peripheral blood. These antibodies were used in immunocytochemistry, in flow cytometry or immunomagnetic selection, e.g. GB17 and GB25 by Cacheux et al. [17], monoclonal M340 by Durrant et al. [18] and FDO46B, FDO161G, FDO101X antibodies (specific for syncytiotrophoblast cells) by Hawes et al. [19]. The markers we used for identification of extravillous uninuclear circulating trophoblast cells are HASH2 and HLA-G.

HASH2

In analogy to the murine gene mammalian Achaete-Scute homologue 2 *(Mash2)*, we cloned, in cooperation with the groups of Marcel Mannens and Peter

Little, in 1997 the human gene HASH2 [20]. *Mash2* is an imprinted gene encoding a basic helix loop helix containing transcription factor, expressed in the murine placenta and essential for placental development. Mice deficient for *Mash2* die at 10 days post coitum due to placental failure as a consequence of deficient spongiotrophoblast formation [21]. The human homologue, designated HASH2 (human Achaete-Scute homologue 2), is expressed in extravillous trophoblast cells and was therefore tested as a marker for circulating trophoblast cells in pregnancy. Using the two-step enrichment procedure consisting of density gradient centrifugation followed by negative selection of CD50-positive maternal leukocytes by magnetic activated cell sorting, we isolated cell fractions enriched for trophoblast cells. In these cell fractions, HASH2-specific mRNA was detected in subsets of cells (fig. 1) by nonradioactive RNA in situ hybridization using an in vitro generated digoxigenin-labeled RNA probe. About 20 cells positive for HASH2 mRNA were detected in 20 ml maternal peripheral blood of pregnant women [22]. Screening of blood of non-pregnant women, processed identically, yielded no positive cells. However, definite proof that HASH2-positive cells identified in the maternal blood are indeed fetal would be obtained by performing X/Y FISH on HASH2-stained cells. Unfortunately, the HASH2-RNA in situ hybridization was not compatible with the FISH procedure currently employed, due to the highly stringent hybridization conditions of the HASH2 probe as a consequence of the extremely high G/C content of the sequence (>70%). Therefore, we developed a new multiparameter in situ procedure consisting of identification of trophoblast cells by immunophenotyping followed by FISH for chromosomal analysis. For this purpose polyclonal HASH2 serum was generated using specific peptides as antigens. However, while this serum did show affinity for recombinant HASH2-GFP, no wild-type protein was detected and therefore this serum was not useful for trophoblast identification (unpublished data). To date we use HLA-G-specific monoclonal antibody G233, characterized as a marker of extravillous trophoblast cells by Loke et al. [23].

HLA-G

The monoclonal antibody G233 stains all populations of extravillous trophoblast cells, interstitial as well as endovascular, whereas an extensive range of adult and fetal tissues tested remains negative [23]. A combined assay was developed for immunocytochemical identification, followed by FISH in order to prove the fetal identity of the stained cells thereby permitting subsequent detection of chromosomal aberrations in these cells. The monoclonal antibody G233 is detected in a three-step detection protocol, using an anti-mouse digoxigenin-labeled secondary antibody and an anti-digoxigenin peroxidase-labeled tertiary antibody (both Fab fragments). The amplification of the signal using AMCA-tyramide results in a strong fluorescently labeled cell. In this combined cytochemical method, immunophenotyping and simultaneous detection of X-chromosome (SpectrumAqua-labeled X-probe), Y-chromosome (Spectrum green-labeled Y-probe), and chromosome 21 (Spectrum orange-labeled probe) can be performed (fig. 1). Using this specific XY-FISH/immunostaining procedure we were able to detect HLA-G-stained XY cells in the blood of a woman carrying a male fetus [24]. In a group of 21 pregnant women (gestational age varying from 7 to 14 weeks) we analyzed the presence of trophoblast cells using this assay. As in situ hybridization is never 100% efficient, we always performed simultaneous X- and Y-chromosome analysis. Only cells with a clear Y-chromosome specific signal and one single X-chromosome specific signal were considered to be male. In 81% of all cases, fetal sex was correctly predicted using HLA-G as a marker for trophoblast cells followed by a FISH using X- and Y-chromosome specific probes [24]. In addition, 47,XY+21 HLA-G-positive cells were detected in a woman carrying a male trisomic fetus. Our conclusion was that it is feasible to use trophoblast cells for noninvasive prenatal screening of chromosomal abnormalities. On average 1 fetal cell was found per milliliter of peripheral blood, which is in good correlation with the data generated by the RNA in situ hybridization assays using HASH2 as a trophoblast specific marker [22] as well as similar to the number of fetal NRBCs identified by other groups at similar gestational periods [15].

The Trophoblast Cell as a Predictor for Preeclampsia

In preeclampsia and intrauterine growth retardation, trophoblast cell invasion is incomplete and defective [10, 11] and as a consequence uterine spiral arteries remain unmodified. In placentae of patients with preeclampsia extravillous trophoblast cells show an abnormal expression profile of specific cell adhesion molecules. Molecules characteristic for the epithelial phenotype, normally downregulated in invading extravillous trophoblast cells, remain expressed in preeclampsia and intrauterine growth retardation. Examples are integrin $\alpha 6\beta 4$ and E-cadherin [10, 11]. At the end of 1998 we started a large longitudinal prospective study including at least 600 pregnant women. The aim of this study is to develop a method for identification of women at risk of developing preeclampsia and/or intrauterine growth retardation by phenotypic analysis of trophoblast cells isolated from maternal peripheral blood in early pregnancy. For this purpose, trophoblast cells isolated from the peripheral circulation will be analyzed for expression of integrin $\alpha 6\beta 4$ in order to detect malfunction of trophoblast cells. This might predict the onset of pregnancy-associated diseases like preeclampsia and/or intrauterine growth retardation, prior to the manifestation of clinical symptoms. Therefore a double, two-color immunocytochemical staining has been developed that permits simultaneous analysis of the fetal identity as well as the phenotypic changes reflecting dysfunction of the trophoblast cells positively identified (fig. 1). This method is currently being optimized for use on clinical samples. If necessary, FISH can subsequently be performed on these cells.

To date more than 500 pregnant women have been included in this study. For all patients, heparin blood samples are obtained on three different occasions, between 7 and 20 weeks of gestation. In addition a detailed echo Doppler analysis is performed at 18–20 weeks of gestation to measure flow parameters in the arteria uterina and arteria umbilicalis. All pregnancies are closely monitored and all relevant clinical and laboratory data are stored in a large database. We are currently performing the first double immunocytochemical stainings on a set of slides stored at –20 °C, containing enriched, fixed and immobilized cells from peripheral blood in early pregnancy.

Plasma DNA

Many similarities exist between the invasion of extra-embryonic cells into the maternal endometrium and invasion of tumor cells. The biological mechanisms of trophoblast invasion might lead to a better understanding of growth and spread of cancer cells [25]. Inversely, the reports of tumor DNA detected in the plasma and serum of cancer patients, prompted Lo et al. [26] to investigate in pregnant women the presence of plasma DNA derived from the fetus. In 1997, they were the first to describe the presence of fetal DNA in maternal plasma and serum during pregnancy. Recently, we provided evidence that not all of this fetal DNA is cell-free, but that part of it is still cell-associated [27]. After density centrifugation of maternal peripheral blood applied for isolation of trophoblast cells, we used the upper part of the gradient containing (diluted) plasma to isolate fetal DNA (fig. 1). Surprisingly, following two centrifugation steps, cells were detected. Moreover, and most interestingly, the majority of these cells were proven to be in apoptosis as shown by their characteristic nuclear morphology and as confirmed by TUNEL assays. This result was confirmed by the ANNEXIN-V assay validating the apoptotic state of these cells and excluding the existence of necrosis (unpublished data). Using the nested specific and sensitive X/Y PCR method described above, we proved that fetal DNA is present in these cell fractions. In addition, by performing FISH, we demonstrated the usefulness of these cells for noninvasive prenatal diagnosis of chromosomal aberrations [27] (fig. 1).

Plasma DNA has been postulated to be a marker of cell death. Increased amounts of fetal DNA may be liberated from apoptotic areas in the placenta. Both in preeclampsia and intrauterine growth retardation increased levels of apoptosis are found [28, 29]. Accordingly, increased levels of plasma DNA are detected in preeclampsia [30]. It is hypothesized that at least part of the fetal DNA in plasma is derived from trophoblast cells. We are currently investigating the phenotype of the fetal apoptotic cells detected in the plasma of pregnant women. If a subset of these fetal apoptotic cells isolated from the plasma of pregnant women is indeed trophoblast cells, these cells may not only be used for noninvasive prenatal diagnosis of numeric chromosomal aberrations, but also for early identification of pregnancies at risk of development of preeclampsia or intrauterine growth restriction. This could be performed either as described above, by phenotypic analysis, although this might be difficult in apoptotic cells, or by quantitative analysis of these cells. However, the major advantage of the use of these cells is that they are rapidly available by simple centrifugation of plasma samples.

Acknowledgements

We thank Prof. Dr. Y.W. Loke, University of Cambridge, for his collaboration and for providing the G233 antibody. We wish to thank Drs M. Haak, T. Medema and E. Zanders from the Prenatal Diagnostics Center for counselling the patients. We thank Vysis Inc. for their collaboration and for providing the CEP probes.

This work is financially supported by the 'Praeventiefonds/ZON' (P28-3022).

References

1 Pijnenborg R: The placental bed. Hypertens Pregnancy 1996;15:17–23.
2 Craven CM, Zhao L, Ward K: Lateral placental growth occurs by trophoblast cell invasion of decidual veins. Placenta 2000;21:160–169.
3 Zhou Y, Fisher SJ, Janatpour M, Genbacev O, Dejana E, Wheelock M, Damsky CH: Human cytotrophoblasts adopt a vascular phenotype as they differentiate. J Clin Invest 1997;99:2139–2151.
4 Kam EP, Gardner L, Loke YW, King A: The role of trophoblast in the physiological change in decidual spiral arteries. Hum Reprod 1999;14:2131–2138.
5 Michel MZ, Khong TY, Clark DA, Beard RW: A morphological and immunological study of human placental bed biopsies in miscarriage. Br J Obstet Gynaecol 1990;97:984–988.
6 Meekins JW, Pijnenborg R, Hanssens M, McFayden IR, van Asshe A: A study of placental bed spiral arteries and trophoblast invasion in normal and severe pre-eclamptic pregnancies. Br J Obstet Gynaecol 1994;101:669–674.
7 King A, Loke YW: Unexplained fetal growth retardation: What is the cause? Arch Dis Child Fetal Neonatal Ed 1994;70:F225–F227.
8 Goldman-Wohl DS, Ariel I, Greenfield C, Hochner-Celnikier D, Cross J, Fisher S: Lack of human leukocyte antigen-G expression in extravillous trophoblasts is associated with pre-eclampsia. Mol Hum Reprod 2000;6:88–95.
9 Aldrich C, Verp MS, Walker MA, Ober C: A null mutation in HLA-G is not associated with preeclampsia or intrauterine growth retardation. J Reprod Immunol 2000;47:41–48.
10 Zhou Y, Damsky CH, Chiu K, Roberts JM, Fisher SJ: Pre-eclampsia is associated with abnormal expression of adhesion molecules by invasive cytotrophoblasts. J Clin Invest 1993;91:950–960.
11 Zhou Y, Damsky CH, Fischer SJ: Pre-eclampsia is associated with failure of human cytotrophoblast to mimic a vascular adhesion phenotype. J Clin Invest 1997;99:2152–2164.
12 Pertl B, Bianchi DW: First trimester prenatal diagnosis: Fetal cells in the maternal circulation. Semin Perinatol 1999;23:393–402.

13 Thomas MR, Tutschek B, Frost A, Rodeck CH, Yazdani N, Craft I Williamson R: The time of appearance and disappearance of fetal DNA from the maternal circulation. Prenatal Diagn 1995; 15:641–646.
14 van Wijk IJ, van Vugt JM, Mulders MA, Können AA, Weima SM, Oudejans CB: Enrichment of fetal trophoblast cells from the maternal peripheral blood followed by detection of fetal deoxyribonucleic acid with a nested X/Y polymerase chain reaction. Am J Obstet Gynecol 1996;174:871–878.
15 Bianchi DW, Williams JM, Sullivan LM, Hanson FW, Klinger KW, Shuber AP: PCR quantitation of fetal cells in maternal blood in normal and aneuploid pregnancies. Am J Hum Genet 1997;61: 822–829.
16 Dingemans AM, Brakenhoff RH, Postmus PE, Giaccone G: Detection of cytokeratin-19 transcripts by reverse transcriptase-polymerase chain reaction in lung cancer cell lines and blood of lung cancer patients. Lab Invest 1997;77:213–220.
17 Cacheux V, Milesi-Fluet C, Tachdjian G, Druart L, Bruch JF, Hsi BL: Detection of 47,XYY trophoblast fetal cells in maternal blood by fluorescence in situ hybridization after using immunomagnetic lymphocyte depletion and flow cytometry sorting. Fetal Diagn Ther 1992;7:190–194.
18 Durrant L, McDowall K, Holmes R, Liu D: Non-invasive prenatal diagnosis by isolation of both trophoblasts and fetal nucleated red blood cells from the peripheral blood of pregnant women. Br J Obstet Gynaecol 1996;103:219–222.
19 Hawes CS, Suskin HA, Petropoulos A, Latham SE, Mueller UW: A morphologic study of trophoblast isolated from peripheral blood of pregnant women. Am J Obstet Gynecol 1994;170:1297–1300.
20 Alders M, Hodges M, Hadjantonakis AK, Postmus J, van Wijk I, Bliek J, de Meulemeester M, Westerveld A, Guillemot F, Oudejans C, Little P, Mannens M: The human Achaete-Scute homologue 2 (ASCL2, HASH2) maps to chromosome 11p15.5, close to IGF2 and is expressed in extravillus trophoblasts. Hum Mol Genet 1997;6:859–867.
21 Guillemot F, Nagy A, Auerbach A, Rossant J, Joyner A: Essential role of Mash2 in extraembryonic development. Nature 1994;371:333–336.
22 van Wijk IJ, van Vugt JMG, Können AAM, Mulders MAM, Florijn WJ, Oudejans CBM: Identification of HASH2 positive extravillus trophoblast cells in the peripheral blood of pregnant women. Trophoblast Res 1998;11:23–33.
23 Loke YW, King A, Burrows T, Gardner L, Bowen S, Hiby S, et al: Evaluation of trophoblast HLA-G antigen with a specific monoclonal antibody. Tissue Antigens 1997;50:135–146.
24 van Wijk IJ, Griffioen S, Tjoa ML, Mulders MAM, van Vugt JMG, Loke YW, Oudejans CBM: HLA-G expression in trophoblast cells circulating in maternal peripheral blood during early pregnancy. Am J Obstet Gynecol 2001;in press.
25 Murray MJ, Lessey BA: Embryo implantation and tumor metastasis: Common pathways of invasion and angiogenesis. Semin Reprod Endocrinol 1999;17:275–290.
26 Lo YMD, Corbetta N, Chamberlain PF, Rai V, Sargent IL, Redman CWG, Wainscoat JS: Presence of fetal DNA in maternal plasma and serum. Lancet 1997;350:485–487.
27 van Wijk IJ, de Hoon AC, Jurhawan R, Tjoa ML, Griffioen S, Mulders MA, van Vugt JM, Oudejans CB: Detection of apoptotic fetal cells in plasma of pregnant women. Clin Chem 2000;46:729–731.
28 Genbacev O, DiFederico E, McMaster M, Fisher SJ: Invasive cytotrophoblast apoptosis in preeclampsia. Hum Reprod. 1999;14:59–66.
29 Axt R, Kordina AC, Meyberg R, Reitnauer K, Mink D, Schmidt W: Immunohistochemical evaluation of apoptosis in placentae from normal and intrauterine growth-restricted pregnancies. Clin Exp Obstet Gynecol 1999;26:195–208.
30 Lo YMD, Leung TN, Tein MSC, Sargent IL, Zhang J, Lau TK, Haines CJ, Redman CWG: Quantitative abnormalities of fetal DNA in maternal serum in preeclampsia. Clin Chem 1999;45:184–188.

I.J. van Wijk, PhD, Molecular Biology Laboratory, Department of Clinical Chemistry, University Hospital Vrije Universiteit, PO BOX 7057, NL–1007 MB Amsterdam (The Netherlands)
Tel. +31 20 4443891, Fax +31 20 4443895, E-Mail ij.vanwijk@azvu.nl

Hahn S, Holzgreve W (eds): Fetal Cells and Fetal DNA in Maternal Blood. New Developments for a New Millennium. 11th Fetal Cell Workshop, Basel, 2000. Basel, Karger, 2001, pp 120–123

Transcervical Sampling

Yousef P. Daryani[a], *Graham H. Barker*[b], *Leonie K. Penna*[a,b], *Michael A. Patton*[a]

Departments of [a] Medical Genetics and [b] Obstetrics and Gynaecology, St George's Hospital Medical School, London, UK

Two methods used to detect cells of fetal origin without an increased risk of miscarriage include the detection of fetal cells from maternal circulation [1] and from transcervical sampling [2]. The former requires the identification of a very small number of fetal cells from the maternal cells, although the fetal cells found may not be a representative of the ongoing pregnancy [3]. An alternative approach is to detect trophoblastic cells from the lower part of the uterus or cervical mucus by means of irrigation or aspiration [4]. Unlike blood samples, operator variability may affect the successful retrieval of fetal cells from transcervical samples. We aimed to show this variability and the difference between transcervical irrigation and cervical mucus aspiration in terms of successful fetal cell detection.

Material and Methods

The subjects were women undergoing first-trimester termination of pregnancy for social reasons who had agreed to participate in this project. Forty-seven transcervical irrigation and 16 endocervical mucus samples were analysed. The material and methods have been previously described in detail [5, 6]. For transcervical irrigation 5–8 ml of sterile normal saline was introduced into the lower segment of uterine cavity via a catheter and gently aspirated. Endocervical mucus was aspirated from the endocervical canal with similar technique. The conventional chromosome analysis, i.e. chorionic villus sampling or amniocentesis, by means of karyotyping or FISH was subsequently performed for actual sexing of unborn babies.

Statistical Analysis

Univariate analysis of the differences between the means was performed by Student's t test. A value of $p < 0.05$ was considered significant.

Results

Of 47 pregnancies 40 fetuses were correctly sexed by transcervical irrigation. The mean and median gestational age as determined by scan or LMP was 10 (range 7–14) weeks. The mean age of patients was 27 years.

There were 25 male fetuses of which 19 were correctly predicted. In the 19 the proportion of Y signal in total cells was between 0.7 and 52.2% (median 4.6%). There were 22 female pregnancies of which 21 were correctly predicted. The false-positive result was obtained in a sample with an 8.3% male signal ratio.

The male signals (XY) were detected in 19 of 25 male fetuses (76%) and 1 of 22 female fetuses (< 5% false-positive). The frequency of cells showing XY signals varied between 0.7 and 52.2% (mean 8.6%) in male predicted pregnancies (n = 19).

Two operators had obtained the samples (operator A had 11 and operator B had 8 samples). The mean Y signal ratio was 13.4% (range 2.25–52.2%) for operator A and 2.1% (range 0.7–5.2%) for operator B (p value for Student's t test = 0.026). This indicates that the number of fetal cells obtained, hence the successful retrieval of fetal cells from transcervical irrigations, was affected by operator variability.

Endocervical mucus was also collected from a group of 16 pregnant women by a similar approach. There were 8 male and 8 female pregnancies. Only 3 of 8 male pregnancies (37.5%) were correctly predicted. No false-positive was reported.

In order to study the discomfort associated with the transcervical irrigation a number of pregnant women were asked to compare the procedure with a cervical smear. Seven women participated. The procedure in these cases was performed without anaesthesia and in the antenatal clinic. Six women who had previously had a cervical smear described it as a similar experience.

Comments

Shettles [7] first reported the presence of trophoblastic cells from the birth canal. Since then more studies have been published with conflicting results (reviewed by Adinolfi [4]). The transcervical retrieval of fetal cells will need to be consistently successful to become an alternative to amniocentesis and chorion villus sampling (CVS). It may be considered as an add-on to the biochemical screening and be offered to pregnant women in their first trimester of pregnancy in the future, if proven safe.

The results presented here indicate that cells of fetal origin can be collected successfully in the first trimester as early as 7 weeks after conception with the accuracy as high as 76%. In this study one false-positive result was observed. This was thought to be due to a male vanishing twin, or placental mosaicism. Although the presence of sperm in the samples has been reported [6], this could not be attributed to the misdiagnosis since sperm has a distinct morphology that can be detected under the microscope; i.e. it is haploid, has a tail and a small cytoplasm. The false-negative results in this study may be due to improper sampling or misplacing of the catheter.

The number of the cells and their different types in each sample are expected to be variable since the technique has not been standardised. Even the same operator may aspirate back a different amount of fluid from one case to another. Although this may be due to the nature of the procedure there might be a need in the future to standardise the procedure so its effectiveness will not be compromised by technical error and operator variability. A specifically designed catheter may improve this situation.

This technique is simple and can be performed in a matter of minutes. It has also been reported to be safe in a study by Rhine et al. [8] with no increase in associated complications of pregnancy such as uterine cramping, vaginal spotting, bleeding, temperature elevation, and miscarriage. In another recent study [9] no difference in the outcome was found when compared to a control group who had only CVS. Our study also suggests that the discomfort associated with the technique is comparable to a cervical smear. However, there are still concerns regarding the safety of the procedure including the risk of miscarriage. There was also a report of a rare complication [10] associated with uterine lavage. In that, a special catheter had been introduced up to 3 cm inside the uterus at 7–8 weeks' gestation. This was clearly an invasive and totally different procedure from that practised in the UK. Nevertheless more detailed evaluations are required as a different approach may lead to a different outcome [11].

This study further proves the presence of cells of fetal origin in the birth canal. It also demonstrates that it might be feasible to use this technique for first-trimester prenatal diagnosis. However, the technique needs to be optimised and the cells require isolation and purification.

References

1 Bianchi DW: Prenatal diagnosis by analysis of fetal cells in maternal blood. J Pediatr 1995;127: 847–856.

2 Adinolfi M, Davies A, Sharif S, Soothill P, Rodeck C: Detection of trisomy 18 and Y-derived sequences in fetal nucleated cells by transcervical flushing. Lancet 1993;342:403–404.

3 Bianchi DW, Zickwolf GK, Weil GJ, Sylvester S, De Maria MA: Male fetal progenitor cells persist in maternal blood for as long as 27 years postpartum. Proc Natl Acad Sci USA 1996;93:705–708.

4 Adinolfi M: Non- or minimally invasive prenatal diagnostic tests on maternal blood samples or transcervical cells. Prenat Diagn 1995;15:889–896.

5 Daryani YP, Penna LK, Patton MA: The detection of cells of fetal origin in cervical washing samples. Am J Hum Genet 1995;57(suppl):277a.

6 Daryani YP, Penna LK, Patton MA: Detection of cells of fetal origin from transcervical irrigations. Prenat Diagn 1997;17:243–248.

7 Shettles LB: Use of the Y chromosome in prenatal sex determination. Nature 1971;230:52–53.

8 Rhine SA, Palmer CG, Thompson JF: A simple alternative to amniocentesis for first trimester prenatal diagnosis. Birth Defects Orig Artic Ser 1977;12:231–247.

9 Rodeck C, Tutschek B, Sherlock J, Kingdom J: Methods for the transcervical collection of fetal cells during the first trimester of pregnancy. Prenat Diagn 1995;15:933–942.

10 Chou MM, Lin SK, Chu Ho ES: Severe limb defects after uterine lavage at 7–8 weeks' gestation, Prenat Diagn 1997;17:77–80.

11 Daryani YP, Barker G, Patton MA: Limb defects as a consequence of CVS or uterine lavage? A need for further documentation of the technique and complications. Prenat. Diagn 1997;17:789–791.

Dr. Yousef Daryani, Department of Medical Genetics, St George's Hospital Medical School
Cranmer Terrace, London SW17 0RE (UK)
Tel. +44 20 8725 5335, Fax +44 20 8725 3444, E-Mail ydaryani@doctors.org.uk

Hahn S, Holzgreve W (eds): Fetal Cells and Fetal DNA in Maternal Blood. New Developments for a New Millennium. 11th Fetal Cell Workshop, Basel, 2000. Basel, Karger, 2001, pp 124–130

Impact of Improved Screening Efficiency on the Future Role of Non-Invasive Testing

Howard Cuckle

Reproductive Epidemiology, University of Leeds, UK

Introduction

Currently Down's syndrome screening is predicated on the fact that prenatal diagnosis of this disorder is only possible through invasive procedures which are expensive and hazardous. The sole aim of screening is to identify women at high enough risk to justify invasive prenatal diagnosis. If a cheap means of diagnosis using maternal blood is eventually found and it is as reliable as amniotic fluid, chorionic villus or fetal blood testing then screening would no longer be needed. However, whilst such an outcome remains the ultimate objective it is prudent to consider whether a relatively expensive not completely accurate non-invasive test would be clinically useful. The answer to this question is dependent on how efficient screening eventually becomes. Efficiency in this context refers to the highest detection rate that can be achieved for a given false-positive rate and can best be estimated by statistical modelling with parameters derived, wherever possible, from a meta-analysis of all published series (see Appendix).

Current Screening Efficiency

Among the many markers of Down's syndrome, seven are widely used in screening: human chorionic gonadotrophin (hCG), the free-β subunit of hCG, α-fetoprotein (AFP), unconjugated oestriol (uE_3), inhibin A, pregnancy-associated plasma protein (PAPP)-A, and nuchal translucency (NT). Levels are usually

Table 1. Seven markers: average level in Down's syndrome

Marker	Operating range, weeks	Source	Down's syndrome	Average MoM	95% CI
NT	11–13	[1]	326	2.02	1.93–2.12
PAPP-A	6–8	[2]	31	0.35	0.25–0.49
	9–11	[2]	197	0.40	0.35–0.46
	12–14	[2]	113	0.62	0.52–0.74
Free β-hCG	<14	[2]	579	1.98	1.83–2.10
	14+	[3]	477	2.30	2.13–2.49
hCG	14+	[3]	850	2.02	1.91–2.13
Inhibin A	13+	[4, 5][1]	570	1.84	1.73–1.95
AFP	<14	[2]	542	0.79	0.75–0.84
	14+	[3]	1,140	0.73	0.71–0.75
uE_3	<14	[2]	226	0.74	0.67–0.82
	14+	[3]	613	0.73	0.70–0.76

[1] Ten studies cited in Lam and Tang [4] and a further study by Renier et al. [5].

expressed as multiples of the normal median (MoM) for the relevant gestation, in order to standardise for both gestational and laboratory differences. Table 1 shows the average level for each marker in Down's syndrome pregnancies. The serum marker levels are based on meta-analysis of non-intervention studies whereas NT is from a single large intervention study. The NT study is the only one whose results have been adjusted for the inevitable over-diagnosis of non-viable cases.

Many different marker combinations are possible but among those used today the most efficient are those using NT together with four serum markers: free β-hCG, AFP, uE_3 and either PAPP-A in the first trimester or inhibin in the second. Table 2 shows that the efficiency of NT alone and the two sets of four serum markers alone are similar, but that it increases substantially when they are combined.

Near-Future Developments

Various extensions to current practice have been suggested which could, in principle, lead to even greater screening efficiency. These involve the use of additional maternal serum or urine markers and maternal DNA polymorphisms. More research is needed before any of them can be turned into practical screening

Table 2. Model-predicted Down's syndrome detection rate for a fixed false-positive rate according to maternal marker combination

Marker combination[1]	Gestation weeks	False-positive rate, %		
		1%	3%	5%
NT alone	11–13	59.5	68.3	72.9
Four early serums	9–11	56.0	70.5	77.1
Four late serums	14+	50.5	64.9	71.8
Four early serums and NT	11 and *12*	77.2	85.4	88.9
NT and four late serums	*13* and 14	77.6	85.5	88.8

Italics indicate the NT values.
[1] Four early serums = free β-hCG, AFP, uE_3 and PAPP-A; four late serums = free β-hCG, AFP, uE_3 and inhibin A.

strategies. But statistical modelling with the latest available parameters, and making appropriate assumptions, is a guide to how much screening efficiency might increase.

Additional Serum Markers

One suggestion is to take two serum samples, at about 10 and 15 weeks gestation, testing the first for PAPP-A and the second for hCG, AFP, uE_3 and inhibin [6]. In addition, all women would have NT at about 12 weeks but neither this nor the PAPP-A result would be acted on until the second-trimester levels are known. Table 3 shows the results for a modified version of this so-called 'integrated test' with the waiting time for the result reduced to 3 weeks and using free β-hCG instead of hCG. Compared with the best results in table 2, an increase in the detection rate of 3.0–4.7% could be achieved, depending on the false-positive rate.

Another approach is to take only one serum sample but increase the number of markers to five, as well as measuring NT. There are several possible candidates for the additional marker but for illustration purposes I will consider only pregnancy-specific β_1-glycoprotein (SP-1) and the 43-kD subunit of placental isoferritin (p43).

Maternal serum SP-1 levels are reduced on average in the first trimester and increased in the second. The average level in a total of 111 published cases at 10–14 weeks gestation is 0.81 MoM and in 379 at 15–22 weeks is 1.47 MoM [7]. Table 3 shows that the use of SP-1 as a fifth first-trimester serum marker would

Table 3. Model-predicted Down's syndrome detection rate for a fixed false-positive rate according to the maternal marker combination

Marker combination[1]	Gestation weeks	False-positive rate, %		
		1%	3%	5%
PAPP-A, NT and four late serums	11, *12* and 14	82.3	89.1	91.9
Four early serums, SP-1 and NT	11 and *12*	78.3	86.4	89.7
NT, four late serums and SP-1	*13* and 14	81.5	88.7	91.6
Four early serums, p43 and NT	11 and *12*	81.5	89.0	92.1
NT, four late serums and p43	*13* and 14	81.1	88.4	91.3
NT, four late serums and urine β-core hCG	*13* and 14	78.1	86.0	89.3
Four early serums, polymorphisms and NT	11 and *12*	78.2	86.3	89.7
NT, polymorphisms and four late serums	*13* and 14	78.5	86.3	89.6

Italics indicate the NT values.

[1] Four early serums = free β-hCG, AFP, uE_3 and PAPP-A; four late serums = free β-hCG, AFP, uE_3 and inhibin A; polymorphisms are 677C→T and 66A→G.

only increase the detection rate by 0.8–1.1% whereas in the second trimester there would be a 2.8–3.9% increase. The first-trimester results would probably be better if samples were taken earlier than 11 weeks gestation as there is a tendency for the average level in Down's syndrome pregnancies to become closer to 1 MoM as the first trimester advances.

Maternal serum p43 has only been measured in a single series of 42 Down's syndrome pregnancies but unlike SP-1 it appears to have a wide operating range [8]. The median level in 19 cases tested before 14 weeks gestation was 1.69 MoM and in 23 cases at 14 weeks or more it was 1.43 MoM. Table 3 shows that p43 as a fifth marker would increase detection by 3.2–4.3 and 2.4–3.5% in the first and second trimester, respectively.

Maternal Urine Markers

The marker that has been most studied so far is the β-core fragment of hCG, its major metabolic product. In the second trimester of pregnancy the mean based on a total of 364 published cases is 3.64 MoM with a 95% confidence interval of 3.11–4.25 [9]. In the first trimester levels are not raised to any great extent.

One possibility is to take a urine sample at the same time as the second-trimester serum sample and test it for β-core hCG. Table 3 shows that this would increase the detection rate by just 0.5%. However, caution is needed in interpreting this result as there is significant heterogeneity between the published studies, so the overall average value in Down's syndrome used in the modelling may be misleading.

Two other urinary hCG species, free β-hCG and hyperglycosylated hCG, are elevated on average in affected pregnancies although fewer cases have been tested than for β-core hCG. In contrast maternal urine total oestrogen and total oestriol levels are reduced on average in Down's syndrome pregnancies. So other possible serum-urine combinations might also be considered.

Maternal Polymorphisms

In two studies the common 677C→T polymorphism in the methylenetetrahydro-folate reductase gene was reported to occur more frequently than usual in the mothers of children with Down's syndrome [10, 11]. One study [10] also found increased frequency of another folate polymorphism, namely 66A→G in the methionine synthase reductase gene. The authors also presented data suggesting that the two polymorphisms act independently. Thus the population can be divided into 3 groups comprising about 4/10, 5/10 and 1/10 of women with relative risks of about 1, 2 and 4, respectively.

Table 3 shows the extent to which routinely testing women for these two polymorphisms and altering the prior risk accordingly would increase screening efficiency. The detection rate would increase by 0.8–1.0% in the first trimester and 0.8–0.9% in the second.

Sequential Screening and Diagnostics

There are financial and practical restraints which might limit the implementation of these newer approaches. But a compromise is possible which leads to a minimal loss of efficiency. This involves sequential testing with multiple cut-off levels. Thus those women having an extremely high risk using a current test would be offered prenatal diagnosis and those with a moderately elevated risk would go on to have further marker or polymorphism determination before deciding on the next step. This already happens to some extent when the NT has been measured but the serum markers have not yet been assayed.

A multi-tiered strategy could also be considered for diagnostic testing. The labour intensive nature of cytogenetics means that the cost of complete karyotyping is high. Multi-coloured FISH [12] and fluorescent PCR [13] are considerably cheaper but in addition to occasional false-negatives for common aneuploidies they fail to detect all other chromosomal re-arrangements. Nevertheless PCR could be regarded as sufficient for women having invasive prenatal diagnosis because of an increased risk of Down's syndrome. If so cytogenetics could be reserved for those at high risk of other chromosomal disorders.

If such an approach were to prove acceptable a natural extension would be to introduce a third tier for women with borderline risk of Down's syndrome. In such women the risk may not be high enough to justify the hazards of invasive prenatal diagnosis yet great enough to warrant the expense of diagnostic testing. A less than perfect test based on circulating fetal DNA may prove to be ideal for these women.

Appendix: Statistical Modelling

Detection and false-positive rates were estimated from the multivariate Gaussian distribution of log marker levels by numerical integration [14] with a Gaussian model of log maternal age. Parameters for the maternal age-specific risks were those in Cuckle et al. [15], for maternal age the mean was 27 and coefficient of variation 16%. Parameters for the current seven markers were derived as follows: NT [1]; free β-hCG, AFP, uE_3 in the first trimester and PAPP-A [2] adjusted to 740 women screened at 9–11 weeks gestation in Leeds and in the second trimester [3] adjusted to 18,000 women screened at 14–19 weeks gestation, and inhibin A [4]. Parameters for SP-1 were: means [7] and standard deviations in the first trimester [16] and in the second trimester correlations with free β-hCG (based on hCG) and AFP [17] and with PAPP-A [18] (based on affected and unaffected pregnancies combined), with uE_3 and inhibin A assumed to be zero. For p43: means, standard deviations and correlations with free β-hCG, AFP and uE_3 [8] with PAPP-A and inhibin A assumed to be zero. For urine β-core hCG: means, standard deviations and correlations with serum free β-hCG, AFP and uE_3 (based only on unaffected pregnancies) [19] with PAPP-A and inhibin A assumed to be zero. For the distribution of polymorphisms in affected and unaffected pregnancies see Hobbs et al. [10].

References

1 Nicolaides KH, Snijders RJM, Cuckle HS: Correct estimation of parameters for ultrasound nuchal translucency screening. Prenat Diagn 1998;18:519–521.

2 Cuckle HS, van Lith JMM: Appropriate biochemical parameters in first trimester screening for Down's syndrome. Prenat Diagn 1999;19:505–512.

3 Cuckle HS: Improved parameters for risk estimation in Down's syndrome screening. Prenat Diagn 1995;15:1057–1065.

4 Lam YH, Tang MHY: Second-trimester maternal serum inhibin-A screening for fetal Down's syndrome in Asian women. Prenat Diagn 1999;19:463–467.

5 Renier MA, Vereecken A, van Herck E, Straetmans D, Ramaeckers P, Buytaert P: Second trimester maternal dimeric inhibin-A in the multiple-marker screening test for Down's syndrome. Hum Reprod 1998;13:744–748.

6 Wald NJ, Watt HC, Hackshaw AK: Integrated screening for Down's syndrome based on tests performed during the first and second trimesters. N Engl J Med 1999;341:461–467.

7 Wald NJ, Kennard A, Hackshaw A, McGuire A: Antenatal screening for Down's syndrome. Health Technol Assess 1998;2:1–112.
8 Moroz C, Maymon R, Jauniaux E, Traub L, Cuckle H: Screening for trisomies 21 and 18 with maternal serum placental isoferritin p43 component. Prenat Diagn 2000;20:395–399.
9 Cuckle H: Antenatal screening for chromosomal and genetic abnormalities; in O'Brien PMS (ed): The Yearbook of Obstetrics and Gynaecology. London, RCOG, 2000, vol 8, pp 74–85.
10 Hobbs CA, Sherman SL, Ping Y, Hopkins SE, Torfs CP, Hine RJ, Pogribna M, Rozen R, James SJ: Polymorphisms in genes involved in folate metabolism as maternal risk factors for Down syndrome. Am J Hum Genet 2000;67:623–630.
11 James SJ, Pogribna M, Pogribny IP, Melnyk S, Hine RJ, Gibson JB, Yi P, Tafoya DL, Swenson DH, Wilson VL, Gaylor DW: Abnormal folate metabolism and mutation in the methylenetetrahydrofolate reductase gene may be maternal risk factors for Down syndrome. Am J Clin Nutr 1999;70: 495–501.
12 Thilaganathan S, Sairam S, Ballard T, Peterson C, Meredith R: Effectiveness of prental chromosomal analysis using multicolour fluorescent in situ hybridisation. Br J Obstet Gynaecol 2000;107: 262–266.
13 Verma L, Macdonald F, Leedham P, McConachie M, Dhanjal S, Hulten M: Rapid and simple prenatal DNA diagnosis of Down's syndrome. Lancet 1998;352:9–12.
14 Royston P, Thompson SG: Model-based screening by risk with application to Down's syndrome. Stat Med 1992;11:257–268.
15 Cuckle HS, Wald NJ, Thompson, SG: Estimating a woman's risk of having a pregnancy associated with Down's syndrome using her age and serum alpha-fetoprotein level. Br J Obstet Gynaecol 1997; 94:387–402.
16 Wald NJ, Watt HC, Norgaard-Pedersen B, Christiansen M: SP1 in pregnancies with Down syndrome in the first trimester of pregnancy. Prenat Diagn 1999;19:517–520.
17 Graham GW, Crossley JA, Aitken DA, Connor JM: Variations in the level of pregnancy-specific β-1-glycoprotein in maternal serum from chromosomally abnormal pregnancies. Prenat Diagn 1992;12:505–512.
18 Bersinger NA, Brizot ML, Johnson A, Snijders RJM, Abbott J, Schneider H, Nicolaides KH: First trimester maternal serum pregnancy-associated plasma protein A and pregnancy-specific 1-glycoprotein in fetal trisomies. Br J Obstet Gynaecol 1994;101:970–974.
19 Cuckle HS, Canick JA, Kellner LH: Collaborative study of maternal urine β-core human chorionic gonadotropin screening for Down's syndrome. Prenat Diagn 1999;19:911–917.

Prof. Howard Cuckle, Reproductive Epidemiology, University of Leeds
26 Clarendon Road, Leeds LS2 9NZ (UK)
Tel. +44 113 233 6771, Fax +44 113 233 6774, E-Mail h.s.cuckle@leeds.ac.uk

Author Index

Subject Index